TWO NEW SCIENCES BY GALILEO

Galileo's birthplace in Pisa

Dialogues Concerning

TWO NEW
SCIENCES

GALILEO GALILEI

TRANSLATED BY
Henry Crew & Alfonso de Salvio

WITH AN INTRODUCTION BY
Antonio Favaro

DOVER PUBLICATIONS, INC., NEW YORK

Published in Canada by General Publishing Company, Ltd.,
30 Lesmill Road, Don Mills, Toronto, Ontario.
Published in the United Kingdom by Constable and
Company, Ltd., 10 Orange Street, London WC 2.

This Dover edition, first published in 1954, is an unabridged
and unaltered republication of the translation by Henry Crew
and Alfonso de Salvio, originally published in 1914 by The
Macmillan Company.

Standard Book Number: 486-60099-8

Library of Congress Catalog Card Number: A54-8849

Manufactured in the United States of America

DOVER PUBLICATIONS, INC.
180 Varick Street
New York, N. Y. 10014

"La Dynamique est la science des forces accélératrices or retardatrices, et des mouvemens variés qu'elles doivent produire. Cette science est due entièrement aux modernes, et Galilée est celui qui en a jeté les premiers fondemens." Lagrange *Mec. Anal.* I. 221.

TRANSLATORS' PREFACE

OR more than a century English speaking students have been placed in the anomalous position of hearing Galileo constantly referred to as the founder of modern physical science, without having any chance to read, in their own language, what Galileo himself has to say. Archimedes has been made available by Heath; Huygens' *Light* has been turned into English by Thompson, while Motte has put the *Principia* of Newton back into the language in which it was conceived. To render the Physics of Galileo also accessible to English and American students is the purpose of the following translation.

The last of the great creators of the Renaissance was not a prophet without honor in his own time; for it was only one group of his country-men that failed to appreciate him. Even during his life time, his *Mechanics* had been rendered into French by one of the leading physicists of the world, Mersenne.

Within twenty-five years after the death of Galileo, his *Dialogues on Astronomy*, and those on *Two New Sciences*, had been done into English by Thomas Salusbury and were worthily printed in two handsome quarto volumes. The *Two New Sciences*, which contains practically all that Galileo has to say on the subject of physics, issued from the English press in 1665.

It is supposed that most of the copies were destroyed in the great London fire which occurred in the year following. We are not aware of any copy in America: even that belonging to the British Museum is an imperfect one.

Again in 1730 the *Two New Sciences* was done into English by Thomas Weston; but this book, now nearly two centuries old, is scarce and expensive. Moreover, the literalness with which this translation was made renders many passages either ambiguous or unintelligible to the modern reader. Other than these two, no English version has been made.

Quite recently an eminent Italian scholar, after spending thirty of the best years of his life upon the subject, has brought to completion the great National Edition of the Works of Galileo. We refer to the twenty superb volumes in which Professor Antonio Favaro of Padua has given a definitive presentation of the labors of the man who created the modern science of physics.

The following rendition includes neither *Le Mechaniche* of Galileo nor his paper *De Motu Accelerato*, since the former of these contains little but the Statics which was current before the time of Galileo, and the latter is essentially included in the Dialogue of the Third Day. Dynamics was the one subject to which under various forms, such as Ballistics, Acoustics, Astronomy, he consistently and persistently devoted his whole life. Into the one volume here translated he seems to have gathered, during his last years, practically all that is of value either to the engineer or the physicist. The historian, the philosopher, and the astronomer will find the other volumes replete with interesting material.

It is hardly necessary to add that we have strictly followed the text of the National Edition—essentially the Elzevir edition of 1638. All comments and annotations have been omitted save here and there a foot-note intended to economize the reader's time. To each of these footnotes has been attached the signature [*Trans.*] in order to preserve the original as nearly intact as possible.

Much of the value of any historical document lies in the language employed, and this is doubly true when one attempts to trace the rise and growth of any set of concepts such as those employed in modern physics. We have therefore made this translation as literal as is consistent with clearness and modernity. In cases where there is any important deviation from this rule, and in the case of many technical terms where there is no deviation from it, we have given the original Italian or Latin phrase in italics enclosed in square brackets. The intention here is to illustrate the great variety of terms employed by the early physicists to describe a single definite idea, and conversely, to illustrate the numerous senses in which, then as now, a single word is used. For the few explanatory English words which are placed in square brackets without italics, the translators alone are responsible. The paging of the National Edition is indicated in square brackets inserted along the median line of the page.

The imperfections of the following pages would have been many more but for the aid of three of our colleagues. Professor D. R. Curtiss was kind enough to assist in the translation of those pages which discuss the nature of Infinity: Professor O. H. Basquin gave valuable help in the rendition of the chapter on Strength of Materials; and Professor O. F. Long cleared up the meaning of a number of Latin phrases.

To Professor A. Favaro of the University of Padua the translators share, with every reader, a feeling of sincere obligation for his Introduction.

H. C.
A. DE S.

EVANSTON, ILLINOIS,
15 *February*, 1914.

INTRODUCTION

WRITING to his faithful friend Elia Diodati, Galileo speaks of the "New Sciences" which he had in mind to print as being "superior to everything else of mine hitherto published"; elsewhere he says "they contain results which I consider the most important of all my studies"; and this opinion which he expressed concerning his own work has been confirmed by posterity: the "New Sciences" are, indeed, the masterpiece of Galileo who at the time when he made the above remarks had spent upon them more than thirty laborious years.

One who wishes to trace the history of this remarkable work will find that the great philosopher laid its foundations during the eighteen best years of his life—those which he spent at Padua. As we learn from his last scholar, Vincenzio Viviani, the numerous results at which Galileo had arrived while in this city, awakened intense admiration in the friends who had witnessed various experiments by means of which he was accustomed to investigate interesting questions in physics. Fra Paolo Sarpi exclaimed: To give us the Science of Motion, God and Nature have joined hands and created the intellect of Galileo. And when the "New Sciences" came from the press one of his foremost pupils, Paolo Aproino, wrote that the volume contained much which he had "already heard from his own lips" during student days at Padua.

Limiting ourselves to only the more important documents which might be cited in support of our statement, it will suffice to mention the letter, written to Guidobaldo del Monte on the 29th of November, 1602, concerning the descent of heavy bodies along the arcs of circles and the chords subtended by them; that to Sarpi, dated 16th of October, 1604, dealing with the free fall of heavy bodies; the letter to Antonio de' Medici on the 11th of February, 1609, in which he states that he has "completed all the theorems and demonstrations pertaining to forces and resistances of beams of various lengths, thicknesses and shapes, proving that they are weaker at the middle than near the ends, that they can carry a greater load when that load is distributed throughout the length of the beam than when concentrated at one point, demonstrating also what shape should be given to a beam in order that it may have the same bending strength at every point," and that he was now engaged "upon some questions dealing with the motion of projectiles"; and finally in the letter to Belisario Vinta, dated 7th of May, 1610, concerning his return from Padua to Florence, he enumerates various pieces of work which were still to be completed, mentioning explicitly three books on an entirely new science dealing with the theory of motion. Although at various times after the return to his native state he devoted considerable thought to the work which, even at that date, he had in mind as is shown by certain fragments which clearly belong to different periods of his life and which have, for the first time, been published in the National Edition; and although these studies were always uppermost in his thought it does not appear that he gave himself seriously to them until after the publication of the *Dialogue* and the completion of that trial which was rightly described as the disgrace of the century. In fact as late as October, 1630, he barely mentions to Aggiunti his discoveries in the theory of motion, and only two years later, in a letter to Marsili concerning the motion of projectiles, he hints at a book nearly ready for publication in which he will treat also of this subject; and only a year after this he writes to Arrighetti that he has in hand a treatise on the resistance of solids.

But the work was given definite form by Galileo during his enforced residence at Siena: in these five months spent quietly with the Archbishop he himself writes that he has completed "a treatise on a new branch of mechanics full of interesting and useful ideas"; so that a few months later he was able to send word to Micanzio that the "work was ready"; as soon as his friends learned of this, they urged its publication. It was, however, no easy matter to print the work of a man already condemned by the Holy Office: and since Galileo could not hope to print it either in Florence or in Rome, he turned to the faithful Micanzio asking him to find out whether this would be possible in Venice, from whence he had received offers to print the *Dialogue on the Principal Systems*, as soon as the news had reached there that he was encountering difficulties. At first everything went smoothly; so that Galileo commenced sending to Micanzio some of the manuscript which was received by the latter with an enthusiasm in which he was second to none of the warmest admirers of the great philosopher. But when Micanzio consulted the Inquisitor, he received the answer that there was an express order prohibiting the printing or reprinting of any work of Galileo, either in Venice or in any other place, *nullo excepto*.

As soon as Galileo received this discouraging news he began to look with more favor upon offers which had come to him from Germany where his friend, and perhaps also his scholar, Giovanni Battista Pieroni, was in the service of the Emperor, as military engineer; consequently Galileo gave to Prince Mattia de' Medici who was just leaving for Germany the first two Dialogues to be handed to Pieroni who was undecided whether to publish them at Vienna or Prague or at some place in Moravia; in the meantime, however, he had obtained permission to print both at Vienna and at Olmütz. But Galileo recognized danger at every point within reach of the long arm of the Court of Rome; hence, availing himself of the opportunity offered by the arrival of Louis Elzevir in Italy in 1636, also of the friendship between the latter and Micanzio, not to mention a visit at Arcetri, he decided to abandon all other plans and entrust to the Dutch publisher the printing of his new work the manuscript of which, although not complete, Elzevir took with him on his return home.

In the course of the year 1637, the printing was finished, and at the beginning of the following year there was lacking only the index, the title-page and the dedication. This last had,

through the good offices of Diodati, been offered to the Count of
Noailles, a former scholar of Galileo at Padua, and since 1634
ambassador of France at Rome, a man who did much to alleviate
the distressing consequences of the celebrated trial; and the
offer was gratefully accepted. The phrasing of the dedication
deserves brief comment. Since Galileo was aware, on the one
hand, of the prohibition against the printing of his works and
since, on the other hand, he did not wish to irritate the Court
of Rome from whose hands he was always hoping for complete
freedom, he pretends in the dedicatory letter (where, probably
through excess of caution, he gives only main outlines) that he
had nothing to do with the printing of his book, asserting that
he will never again publish any of his researches, and will at
most distribute here and there a manuscript copy. He even
expresses great surprise that his new Dialogues have fallen into
the hands of the Elzevirs and were soon to be published; so
that, having been asked to write a dedication, he could think of
no man more worthy who could also on this occasion defend
him against his enemies.

As to the title which reads: *Discourses and Mathematical
Demonstrations concerning Two New Sciences pertaining to Mechanics and Local Motions*, this only is known, namely, that the
title is not the one which Galileo had devised and suggested; in
fact he protested against the publishers taking the liberty of
changing it and substituting "a low and common title for the
noble and dignified one carried upon the title-page."

In reprinting this work in the National Edition, I have followed the Leyden text of 1638 faithfully but not slavishly, because I wished to utilize the large amount of manuscript material which has come down to us, for the purpose of correcting
a considerable number of errors in this first edition, and also
for the sake of inserting certain additions desired by the author
himself. In the Leyden Edition, the four Dialogues are followed
by an "*Appendix containing some theorems and their proofs, dealing with centers of gravity of solid bodies, written by the same
Author at an earlier date*," which has no immediate connection
with the subjects treated in the Dialogues; these theorems were
found by Galileo, as he tells us, "at the age of twenty-two and

after two years study of geometry" and were here inserted only
to save them from oblivion.

But it was not the intention of Galileo that the *Dialogues
on the New Sciences* should contain only the four Days and the
above-mentioned appendix which constitute the Leyden Edition; while, on the one hand, the Elzevirs were hastening the
printing and striving to complete it at the earliest possible date,
Galileo, on the other hand, kept on speaking of another Day,
besides the four, thus embarrassing and perplexing the printers.
From the correspondence which went on between author and
publisher, it appears that this Fifth Day was to have treated
"of the force of percussion and the use of the catenary"; but
as the typographical work approached completion, the printer
became anxious for the book to issue from the press without
further delay; and thus it came to pass that the *Discorsi e
Dimostrazioni* appeared containing only the four Days and the
Appendix, in spite of the fact that in April, 1638, Galileo had
plunged more deeply than ever "into the profound question of
percussion" and "had almost reached a complete solution."

The "New Sciences" now appear in an edition following the
text which I, after the most careful and devoted study, determined upon for the National Edition. It appears also in that
language in which, above all others, I have desired to see it. In
this translation, the last and ripest work of the great philosopher
makes its first appearance in the New World: if toward this
important result I may hope to have contributed in some measure I shall feel amply rewarded for having given to this field of
research the best years of my life.

ANTONIO FAVARO.

UNIVERSITY OF PADUA,
  27th of October, 1913.

DISCORSI
E
DIMOSTRAZIONI
MATEMATICHE,

intorno à due nuoue scienze

Attenenti alla

MECANICA & i MOVIMENTI LOCALI;

del Signor

GALILEO GALILEI LINCEO,

Filosofo e Matematico primario del Serenissimo
Grand Duca di Toscana.

Con vna Appendice del centro di grauità d'alcuni Solidi.

IN LEIDA,

Appresso gli Elsevirii. M. D. C. XXXVIII.

TO THE MOST ILLUSTRIOUS LORD
COUNT OF NOAILLES

Counsellor of his Most Christian Majesty, Knight of the Order
of the Holy Ghost, Field Marshal and Commander,
Seneschal and Governor of Rouergue, and His
Majesty's Lieutenant in Auvergne, my
Lord and Worshipful Patron

MOST ILLUSTRIOUS LORD:—
In the pleasure which you derive from the possession of this work of mine I recognize your Lordship's magnanimity. The disappointment and discouragement I have felt over the ill-fortune which has followed my other books are already known to you. Indeed, I had decided not to publish any more of my work. And yet in order to save it from complete oblivion, it seemed to me wise to leave a manuscript copy in some place where it would be available at least to those who follow intelligently the subjects which I have treated. Accordingly I chose first to place my work in your Lordship's hands, asking no more worthy depository, and believing that, on account of your affection for me, you would have at heart the preservation of my·studies and labors. Therefore, when you were returning home from your mission to Rome, I came to pay my respects in person as I had already done many times before by letter. At this meeting I presented to your Lordship a copy of these two works which at that time I happened to have ready. In the gracious reception which you gave these I found assurance
of

of their preservation. The fact of your carrying them to France and showing them to friends of yours who are skilled in these sciences gave evidence that my silence was not to be interpreted as complete idleness. A little later, just as I was on the point of

[44]

sending other copies to Germany, Flanders, England, Spain and possibly to some places in Italy, I was notified by the Elzevirs that they had these works of mine in press and that I ought to decide upon a dedication and send them a reply at once. This sudden and unexpected news led me to think that the eagerness of your Lordship to revive and spread my name by passing these works on to various friends was the real cause of their falling into the hands of printers who, because they had already published other works of mine, now wished to honor me with a beautiful and ornate edition of this work. But these writings of mine must have received additional value from the criticism of so excellent a judge as your Lordship, who by the union of many virtues has won the admiration of all. Your desire to enlarge the renown of my work shows your unparalleled generosity and your zeal for the public welfare which you thought would thus be promoted. Under these circumstances it is eminently fitting that I should, in unmistakable terms, gratefully acknowledge this generosity on the part of your Lordship, who has given to my fame wings that have carried it into regions more distant than I had dared to hope. It is, therefore, proper that I dedicate to your Lordship this child of my brain. To this course I am constrained not only by the weight of obligation under which you have placed me, but also, if I may so speak, by the interest which I have in securing your Lordship as the defender of my reputation against adversaries who may attack it while I remain under your protection.

And now, advancing under your banner, I pay my respects to you by wishing that you may be rewarded for these kindnesses by the achievement of the highest happiness and greatness.

I am your Lordship's

Most devoted Servant,

GALILEO GALILEI.

Arcetri, 6 March, 1638.

THE PUBLISHER TO THE READER

 INCE society is held together by the mutual services which men render one to another, and since to this end the arts and sciences have largely contributed, investigations in these fields have always been held in great esteem and have been highly regarded by our wise forefathers. The larger the utility and excellence of the inventions, the greater has been the honor and praise bestowed upon the inventors. Indeed, men have even deified them and have united in the attempt to perpetuate the memory of their benefactors by the bestowal of this supreme honor.

Praise and admiration are likewise due to those clever intellects who, confining their attention to the known, have discovered and corrected fallacies and errors in many and many a proposition enunciated by men of distinction and accepted for ages as fact. Although these men have only pointed out falsehood and have not replaced it by truth, they are nevertheless worthy of commendation when we consider the well-known difficulty of discovering fact, a difficulty which led the prince of orators to exclaim: *Utinam tam facile possem vera reperire, quam falsa convincere.** And indeed, these latest centuries merit this praise because it is during them that the arts and sciences, discovered by the ancients, have been reduced to so great and constantly increasing perfection through the investigations and experiments of clear-seeing minds. This development is particularly evident in the case of the mathematical sciences. Here, without mentioning various men who have achieved success, we must without hesitation and with the

* Cicero. *de Natura Deorum*, I, 91. [*Trans.*]

unanimous approval of scholars assign the first place to Galileo Galilei, Member of the Academy of the Lincei. This he deserves not only because he has effectively demonstrated fallacies in many of our current conclusions, as is amply shown by his published works, but also because by means of the telescope (invented in this country but greatly perfected by him) he has discovered the four satellites of Jupiter, has shown us the true character of the Milky Way, and has made us acquainted with spots on the Sun, with the rough and cloudy portions of the lunar surface, with the threefold nature of Saturn, with the phases of Venus and with the physical character of comets. These matters were entirely unknown to the ancient astronomers and philosophers; so that we may truly say that he has restored to the world the science of astronomy and has presented it in a new light.

Remembering that the wisdom and power and goodness of the Creator are nowhere exhibited so well as in the heavens and celestial bodies, we can easily recognize the great merit of him who has brought these bodies to our knowledge and has, in spite of their almost infinite distance, rendered them easily visible. For, according to the common saying, sight can teach more and with greater certainty in a single day than can precept even though repeated a thousand times; or, as another says, intuitive knowledge keeps pace with accurate definition.

But the divine and natural gifts of this man are shown to best advantage in the present work where he is seen to have discovered, though not without many labors and long vigils, two entirely new sciences and to have demonstrated them in a rigid, that is, geometric, manner: and what is even more re-markable in this work is the fact that one of the two sciences deals with a subject of never-ending interest, perhaps the most important in nature, one which has engaged the minds of all the great philosophers and one concerning which an extraordinary number of books have been written. I refer to motion [*moto locale*], a phenomenon exhibiting very many wonderful proper-ties, none of which has hitherto been discovered or demonstrated by any one. The other science which he has also developed from
its

its very foundations deals with the resistance which solid bodies offer to fracture by external forces [*per violenza*], a subject of great utility, especially in the sciences and mechanical arts, and one also abounding in properties and theorems not hitherto observed.

In this volume one finds the first treatment of these two sciences, full of propositions to which, as time goes on, able thinkers will add many more; also by means of a large number of clear demonstrations the author points the way to many other theorems as will be readily seen and understood by all intelligent readers.

TABLE OF CONTENTS

TWO NEW SCIENCES BY GALILEO

FIRST DAY

INTERLOCUTORS: SALVIATI, SA-GREDO AND SIMPLICIO

ALV. The constant activity which you Venetians display in your famous arsenal suggests to the studious mind a large field for investigation, especially that part of the work which involves mechanics; for in this department all types of instruments and machines are constantly being constructed by many artisans, among whom there must be some who, partly by inherited experience and partly by their own observations, have become highly expert and clever in explanation.

SAGR. You are quite right. Indeed, I myself, being curious by nature, frequently visit this place for the mere pleasure of observing the work of those who, on account of their superiority over other artisans, we call "first rank men." Conference with them has often helped me in the investigation of certain effects including not only those which are striking, but also those which are recondite and almost incredible. At times also I have been put to confusion and driven to despair of ever explaining something for which I could not account, but which my senses told me to be true. And notwithstanding the fact that what the old man told us a little while ago is proverbial and commonly accepted, yet it seemed to me altogether false, like many another saying which is current among the ignorant; for I think they introduce these expressions in order to give the appearance of knowing something about matters which they do not understand.

Salv.

[50]

SALV. You refer, perhaps, to that last remark of his when we asked the reason why they employed stocks, scaffolding and bracing of larger dimensions for launching a big vessel than they do for a small one; and he answered that they did this in order to avoid the danger of the ship parting under its own heavy weight [*vasta mole*], a danger to which small boats are not subject?

SAGR. Yes, that is what I mean; and I refer especially to his last assertion which I have always regarded as a false, though current, opinion; namely, that in speaking of these and other similar machines one cannot argue from the small to the large, because many devices which succeed on a small scale do not work on a large scale. Now, since mechanics has its foundation in geometry, where mere size cuts no figure, I do not see that the properties of circles, triangles, cylinders, cones and other solid figures will change with their size. If, therefore, a large machine be constructed in such a way that its parts bear to one another the same ratio as in a smaller one, and if the smaller is sufficiently strong for the purpose for which it was designed, I do not see why the larger also should not be able to withstand any severe and destructive tests to which it may be subjected.

SALV. The common opinion is here absolutely wrong. Indeed, it is so far wrong that precisely the opposite is true, namely, that many machines can be constructed even more perfectly on a large scale than on a small; thus, for instance, a clock which indicates and strikes the hour can be made more accurate on a large scale than on a small. There are some intelligent people who maintain this same opinion, but on more reasonable grounds, when they cut loose from geometry and argue that the better performance of the large machine is owing to the imperfections and variations of the material. Here I trust you will not charge

[51]

me with arrogance if I say that imperfections in the material, even those which are great enough to invalidate the clearest mathematical proof, are not sufficient to explain the deviations observed between machines in the concrete and in the abstract. Yet I shall say it and will affirm that, even if the imperfections

did

did not exist and matter were absolutely perfect, unalterable and free from all accidental variations, still the mere fact that it is matter makes the larger machine, built of the same material and in the same proportion as the smaller, correspond with exactness to the smaller in every respect except that it will not be so strong or so resistant against violent treatment; the larger the machine, the greater its weakness. Since I assume matter to be unchangeable and always the same, it is clear that we are no less able to treat this constant and invariable property in a rigid manner than if it belonged to simple and pure mathematics. Therefore, Sagredo, you would do well to change the opinion which you, and perhaps also many other students of mechanics, have entertained concerning the ability of machines and structures to resist external disturbances, thinking that when they are built of the same material and maintain the same ratio between parts, they are able equally, or rather proportionally, to resist or yield to such external disturbances and blows. For we can demonstrate by geometry that the large machine is not proportionately stronger than the small. Finally, we may say that, for every machine and structure, whether artificial or natural, there is set a necessary limit beyond which neither art nor nature can pass; it is here understood, of course, that the material is the same and the proportion preserved.

Sagr. My brain already reels. My mind, like a cloud momentarily illuminated by a lightning-flash, is for an instant filled with an unusual light, which now beckons to me and which now suddenly mingles and obscures strange, crude ideas. From what you have said it appears to me impossible to build two similar structures of the same material, but of different sizes and have them proportionately strong; and if this were so, it would

[52]

not be possible to find two single poles made of the same wood which shall be alike in strength and resistance but unlike in size.

Salv. So it is, Sagredo. And to make sure that we understand each other, I say that if we take a wooden rod of a certain length and size, fitted, say, into a wall at right angles, i. e.,

parallel

parallel to the horizon, it may be reduced to such a length that it will just support itself; so that if a hair's breadth be added to its length it will break under its own weight and will be the only rod of the kind in the world.* Thus if, for instance, its length be a hundred times its breadth, you will not be able to find another rod whose length is also a hundred times its breadth and which, like the former, is just able to sustain its own weight and no more: all the larger ones will break while all the shorter ones will be strong enough to support something more than their own weight. And this which I have said about the ability to support itself must be understood to apply also to other tests; so that if a piece of scantling [*corrente*] will carry the weight of ten similar to itself, a beam [*trave*] having the same proportions will not be able to support ten similar beams.

Please observe, gentlemen, how facts which at first seem improbable will, even on scant explanation, drop the cloak which has hidden them and stand forth in naked and simple beauty. Who does not know that a horse falling from a height of three or four cubits will break his bones, while a dog falling from the same height or a cat from a height of eight or ten cubits will suffer no injury? Equally harmless would be the fall of a grasshopper from a tower or the fall of an ant from the distance of the moon. Do not children fall with impunity from heights which would cost their elders a broken leg or perhaps a fractured skull? And just as smaller animals are proportionately stronger and more robust than the larger, so also smaller plants are able to stand up better than larger. I am certain you both know that an oak two hundred cubits [*braccia*] high would not be able to sustain its own branches if they were distributed as in a tree of ordinary size; and that nature cannot produce a horse as large as twenty ordinary horses or a giant ten times taller than an

[53]

ordinary man unless by miracle or by greatly altering the proportions of his limbs and especially of his bones, which would have to be considerably enlarged over the ordinary. Likewise the current belief that, in the case of artificial machines the very

* The author here apparently means that the solution is unique.
[*Trans.*]

large and the small are equally feasible and lasting is a manifest error. Thus, for example, a small obelisk or column or other solid figure can certainly be laid down or set up without danger of breaking, while the very large ones will go to pieces under the slightest provocation, and that purely on account of their own weight. And here I must relate a circumstance which is worthy of your attention as indeed are all events which happen contrary to expectation, especially when a precautionary measure turns out to be a cause of disaster. A large marble column was laid out so that its two ends rested each upon a piece of beam; a little later it occurred to a mechanic that, in order to be doubly sure of its not breaking in the middle by its own weight, it would be wise to lay a third support midway; this seemed to all an excellent idea; but the sequel showed that it was quite the opposite, for not many months passed before the column was found cracked and broken exactly above the new middle support.

SIMP. A very remarkable and thoroughly unexpected accident, especially if caused by placing that new support in the middle.

SALV. Surely this is the explanation, and the moment the cause is known our surprise vanishes; for when the two pieces of the column were placed on level ground it was observed that one of the end beams had, after a long while, become decayed and sunken, but that the middle one remained hard and strong, thus causing one half of the column to project in the air without any support. Under these circumstances the body therefore behaved differently from what it would have done if supported only upon the first beams; because no matter how much they might have sunken the column would have gone with them. This is an accident which could not possibly have happened to a small column, even though made of the same stone and having a length corresponding to its thickness, i. e., preserving the ratio between thickness and length found in the large pillar.

[54]

SAGR. I am quite convinced of the facts of the case, but I do not understand why the strength and resistance are not multiplied in the same proportion as the material; and I am the more
puzzled

puzzled because, on the contrary, I have noticed in other cases that the strength and resistance against breaking increase in a larger ratio than the amount of material. Thus, for instance, if two nails be driven into a wall, the one which is twice as big as the other will support not only twice as much weight as the other, but three or four times as much.

SALV. Indeed you will not be far wrong if you say eight times as much; nor does this phenomenon contradict the other even though in appearance they seem so different.

SAGR. Will you not then, Salviati, remove these difficulties and clear away these obscurities if possible: for I imagine that this problem of resistance opens up a field of beautiful and useful ideas; and if you are pleased to make this the subject of to-day's discourse you will place Simplicio and me under many obligations.

SALV. I am at your service if only I can call to mind what I learned from our Academician * who had thought much upon this subject and according to his custom had demonstrated everything by geometrical methods so that one might fairly call this a new science. For, although some of his conclusions had been reached by others, first of all by Aristotle, these are not the most beautiful and, what is more important, they had not been proven in a rigid manner from fundamental principles. Now, since I wish to convince you by demonstrative reasoning rather than to persuade you by mere probabilities, I shall suppose that you are familiar with present-day mechanics so far as it is needed in our discussion. First of all it is necessary to consider what happens when a piece of wood or any other solid which coheres firmly is broken; for this is the fundamental fact, involving the first and simple principle which we must take for granted as well known.

To grasp this more clearly, imagine a cylinder or prism, AB, made of wood or other solid coherent material. Fasten the upper end, A, so that the cylinder hangs vertically. To the lower end, B, attach the weight C. It is clear that however great they may be, the tenacity and coherence [*tenacità e*

* I. e. Galileo: The author frequently refers to himself under this name. [*Trans.*]

[55]

coerenza] between the parts of this solid, so long as they are not infinite, can be overcome by the pull of the weight C, a weight which can be increased indefinitely until finally the solid breaks like a rope. And as in the case of the rope whose strength we know to be derived from a multitude of hemp threads which compose it, so in the case of the wood, we observe its fibres and filaments run lengthwise and render it much stronger than a hemp rope of the same thickness. But in the case of a stone or metallic cylinder where the coherence seems to be still greater the cement which holds the parts together must be something other than filaments and fibres; and yet even this can be broken by a strong pull.

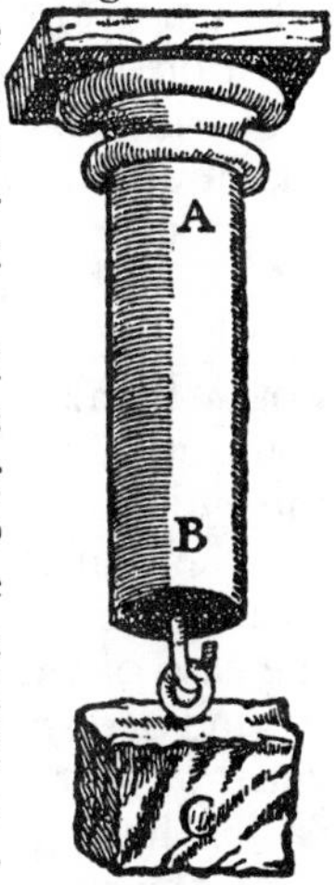

Simp. If this matter be as you say I can well understand that the fibres of the wood, being as long as the piece of wood itself, render it strong and resistant against large forces tending to break it. But how can one make a rope one hundred cubits long out of hempen fibres which are not more than two or three cubits long, and still give it so much strength? Besides, I should be glad to hear your opinion as to the manner in which the parts of metal, stone, and other materials not showing a filamentous structure are

Fig. I

put together; for, if I mistake not, they exhibit even greater tenacity.

Salv. To solve the problems which you raise it will be necessary to make a digression into subjects which have little bearing upon our present purpose.

Sagr. But if, by digressions, we can reach new truth, what harm is there in making one now, so that we may not lose this knowledge, remembering that such an opportunity, once omitted, may not return; remembering also that we are not tied down to a fixed and brief method but that we meet solely for our own entertainment? Indeed, who knows but that we may thus

[56]

frequently

frequently discover something more interesting and beautiful than the solution originally sought? I beg of you, therefore, to grant the request of Simplicio, which is also mine; for I am no less curious and desirous than he to learn what is the binding material which holds together the parts of solids so that they can scarcely be separated. This information is also needed to understand the coherence of the parts of fibres themselves of which some solids are built up.

SALV. I am at your service, since you desire it. The first question is, How are fibres, each not more than two or three cubits in length, so tightly bound together in the case of a rope one hundred cubits long that great force [*violenza*] is required to break it?

Now tell me, Simplicio, can you not hold a hempen fibre so tightly between your fingers that I, pulling by the other end, would break it before drawing it away from you? Certainly you can. And now when the fibres of hemp are held not only at the ends, but are grasped by the surrounding medium through-out their entire length is it not manifestly more difficult to tear them loose from what holds them than to break them? But in the case of the rope the very act of twisting causes the threads to bind one another in such a way that when the rope is stretched with a great force the fibres break rather than separate from each other.

At the point where a rope parts the fibres are, as everyone knows, very short, nothing like a cubit long, as they would be if the parting of the rope occurred, not by the breaking of the filaments, but by their slipping one over the other.

SAGR. In confirmation of this it may be remarked that ropes sometimes break not by a lengthwise pull but by excessive twisting. This, it seems to me, is a conclusive argument because the threads bind one another so tightly that the compressing fibres do not permit those which are compressed to lengthen the spirals even that little bit by which it is necessary for them to lengthen in order to surround the rope which, on twisting, grows shorter and thicker.

SALV. You are quite right. Now see how one fact suggests
another

another. The thread held between the fingers does not yield
[57]
to one who wishes to draw it away even when pulled with considerable force, but resists because it is held back by a double
compression, seeing that the upper finger presses against the
lower as hard as the lower against the upper. Now, if we could
retain only one of these pressures there is no doubt that only
half the original resistance would remain; but since we are
not able, by lifting, say, the upper finger, to remove one of
these pressures without also removing the other, it becomes
necessary to preserve one of them by means of a new device
which causes the thread to press itself against the finger or
against some other solid body upon which it rests; and thus it is
brought about that the very force which pulls
it in order to snatch it away compresses it
more and more as the pull increases. This
is accomplished by wrapping the thread
around the solid in the manner of a spiral;
and will be better understood by means of a
figure. Let AB and CD be two cylinders between which is stretched the thread EF: and
for the sake of greater clearness we will imagine it to be a small cord. If these two
cylinders be pressed strongly together, the
cord EF, when drawn by the end F, will undoubtedly stand a considerable pull before it
slips between the two compressing solids.
But if we remove one of these cylinders the
cord, though remaining in contact with the
other, will not thereby be prevented from
slipping freely. On the other hand, if one
holds the cord loosely against the top of the
cylinder A, winds it in the spiral form AFLOTR, and then
pulls it by the end R, it is evident that the cord will begin to
bind the cylinder; the greater the number of spirals the more
tightly will the cord be pressed against the cylinder by any
given pull. Thus as the number of turns increases, the line of
contact

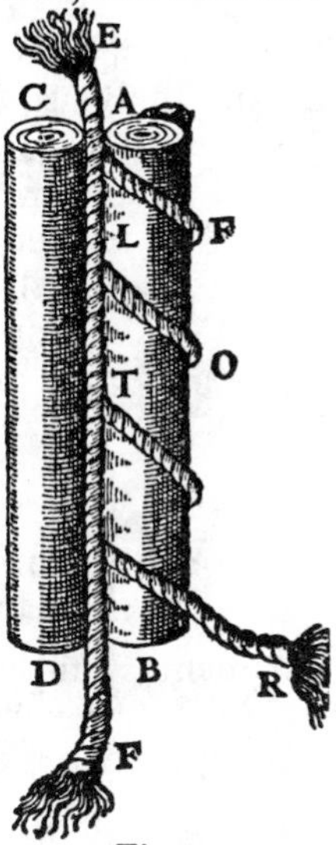

Fig. 2

contact becomes longer and in consequence more resistant; so that the cord slips and yields to the tractive force with increasing difficulty.

[58]

Is it not clear that this is precisely the kind of resistance which one meets in the case of a thick hemp rope where the fibres form thousands and thousands of similar spirals? And, indeed, the binding effect of these turns is so great that a few short rushes woven together into a few interlacing spirals form one of the strongest of ropes which I believe they call pack rope [susta].

SAGR. What you say has cleared up two points which I did not previously understand. One fact is how two, or at most three, turns of a rope around the axle of a windlass cannot only hold it fast, but can also prevent it from slipping when pulled by the immense force of the weight [forza del peso] which it sustains; and moreover how, by turning the windlass, this same axle, by mere friction of the rope around it, can wind up and lift huge stones while a mere boy is able to handle the slack of the rope. The other fact has to do with a simple but clever device, invented by a young kinsman of mine, for the purpose of descending from a window by means of a rope without lacerating the palms of his hands, as had happened to him shortly before and greatly to his discomfort. A small sketch will make this clear. He took a wooden cylinder, AB, about as thick as a walking stick and about one span long: on this he cut a spiral channel of about one turn and a half, and large enough to just receive the rope which he wished to use. Having introduced the rope at the end A and led it out again at the end B, he enclosed both the cylinder and the rope in a case of wood or tin, hinged along the side so that it could be easily opened and closed. After he had fastened the rope to a firm support above, he could, on grasping and squeezing the case with both hands, hang by his arms. The pressure on the rope, lying between the case and the cylinder, was such that he could, at will, either grasp the case

more

more tightly and hold himself from slipping, or slacken his hold and descend as slowly as he wished.

[59]

Salv. A truly ingenious device! I feel, however, that for a complete explanation other considerations might well enter; yet I must not now digress upon this particular topic since you are waiting to hear what I think about the breaking strength of other materials which, unlike ropes and most woods, do not show a filamentous structure. The coherence of these bodies is, in my estimation, produced by other causes which may be grouped under two heads. One is that much-talked-of repugnance which nature exhibits towards a vacuum; but this horror of a vacuum not being sufficient, it is necessary to introduce another cause in the form of a gluey or viscous substance which binds firmly together the component parts of the body.

First I shall speak of the vacuum, demonstrating by definite experiment the quality and quantity of its force [*virtù*]. If you take two highly polished and smooth plates of marble, metal, or glass and place them face to face, one will slide over the other with the greatest ease, showing conclusively that there is nothing of a viscous nature between them. But when you attempt to separate them and keep them at a constant distance apart, you find the plates exhibit such a repugnance to separation that the upper one will carry the lower one with it and keep it lifted indefinitely, even when the latter is big and heavy.

This experiment shows the aversion of nature for empty space, even during the brief moment required for the outside air to rush in and fill up the region between the two plates. It is also observed that if two plates are not thoroughly polished, their contact is imperfect so that when you attempt to separate them slowly the only resistance offered is that of weight; if, however, the pull be sudden, then the lower plate rises, but quickly falls back, having followed the upper plate only for that very short interval of time required for the expansion of the small amount of air remaining between the plates, in consequence of their not fitting, and for the entrance of the surrounding air. This resistance which is exhibited between the two

plates

plates is doubtless likewise present between the parts of a solid, and enters, at least in part, as a concomitant cause of their coherence.

[60]

SAGR. Allow me to interrupt you for a moment, please; for I want to speak of something which just occurs to me, namely, when I see how the lower plate follows the upper one and how rapidly it is lifted, I feel sure that, contrary to the opinion of many philosophers, including perhaps even Aristotle himself, motion in a vacuum is not instantaneous. If this were so the two plates mentioned above would separate without any resistance whatever, seeing that the same instant of time would suffice for their separation and for the surrounding medium to rush in and fill the vacuum between them. The fact that the lower plate follows the upper one allows us to infer, not only that motion in a vacuum is not instantaneous, but also that, between the two plates, a vacuum really exists, at least for a very short time, sufficient to allow the surrounding medium to rush in and fill the vacuum; for if there were no vacuum there would be no need of any motion in the medium. One must admit then that a vacuum is sometimes produced by violent motion [*violenza*] or contrary to the laws of nature, (although in my opinion nothing occurs contrary to nature except the impossible, and that never occurs).

But here another difficulty arises. While experiment convinces me of the correctness of this conclusion, my mind is not entirely satisfied as to the cause to which this effect is to be attributed. For the separation of the plates precedes the formation of the vacuum which is produced as a consequence of this separation; and since it appears to me that, in the order of nature, the cause must precede the effect, even though it appears to follow in point of time, and since every positive effect must have a positive cause, I do not see how the adhesion of two plates and their resistance to separation—actual facts—can be referred to a vacuum as cause when this vacuum is yet to follow. According to the infallible maxim of the Philosopher, the non-existent can produce no effect.

Simp.

SIMP. Seeing that you accept this axiom of Aristotle, I hardly think you will reject another excellent and reliable maxim of his, namely, Nature undertakes only that which happens without resistance; and in this saying, it appears to me, you will find the solution of your difficulty. Since nature abhors a vacuum, she prevents that from which a vacuum would follow as a necessary consequence. Thus it happens that nature prevents the separation of the two plates.

[61]

SAGR. Now admitting that what Simplicio says is an adequate solution of my difficulty, it seems to me, if I may be allowed to resume my former argument, that this very resistance to a vacuum ought to be sufficient to hold together the parts either of stone or of metal or the parts of any other solid which is knit together more strongly and which is more resistant to separation. If for one effect there be only one cause, or if, more being assigned, they can be reduced to one, then why is not this vacuum which really exists a sufficient cause for all kinds of resistance?

SALV. I do not wish just now to enter this discussion as to whether the vacuum alone is sufficient to hold together the separate parts of a solid body; but I assure you that the vacuum which acts as a sufficient cause in the case of the two plates is not alone sufficient to bind together the parts of a solid cylinder of marble or metal which, when pulled violently, separates and divides. And now if I find a method of distinguishing this well known resistance, depending upon the vacuum, from every other kind which might increase the coherence, and if I show you that the aforesaid resistance alone is not nearly sufficient for such an effect, will you not grant that we are bound to introduce another cause? Help him, Simplicio, since he does not know what reply to make.

SIMP. Surely, Sagredo's hesitation must be owing to another reason, for there can be no doubt concerning a conclusion which is at once so clear and logical.

SAGR. You have guessed rightly, Simplicio. I was wondering whether, if a million of gold each year from Spain were not sufficient to pay the army, it might not be necessary to
make

make provision other than small coin for the pay of the soldiers.*

But go ahead, Salviati; assume that I admit your conclusion and show us your method of separating the action of the vacuum from other causes; and by measuring it show us how it is not sufficient to produce the effect in question.

SALV. Your good angel assist you. I will tell you how to separate the force of the vacuum from the others, and afterwards how to measure it. For this purpose let us consider a continuous substance whose parts lack all resistance to separation except that derived from a vacuum, such as is the case with water, a fact fully demonstrated by our Academician in one of his treatises. Whenever a cylinder of water is subjected to a pull and

[62]

offers a resistance to the separation of its parts this can be attrib-

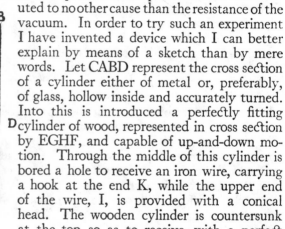

Fig. 4

uted to no other cause than the resistance of the vacuum. In order to try such an experiment I have invented a device which I can better explain by means of a sketch than by mere words. Let CABD represent the cross section of a cylinder either of metal or, preferably, of glass, hollow inside and accurately turned. Into this is introduced a perfectly fitting cylinder of wood, represented in cross section by EGHF, and capable of up-and-down motion. Through the middle of this cylinder is bored a hole to receive an iron wire, carrying a hook at the end K, while the upper end of the wire, I, is provided with a conical head. The wooden cylinder is countersunk at the top so as to receive, with a perfect fit, the conical head I of the wire, IK, when pulled down by the end K.

Now insert the wooden cylinder EH in the hollow cylinder AD, so as not to touch the upper end of the latter but to leave free a space of two or three finger-breadths; this space is to be filled

* The bearing of this remark becomes clear on reading what Salviati says on p. 18 below. [*Trans.*]

with water by holding the vessel with the mouth CD upwards, pushing down on the stopper EH, and at the same time keeping the conical head of the wire, I, away from the hollow portion of the wooden cylinder. The air is thus allowed to escape alongside the iron wire (which does not make a close fit) as soon as one presses down on the wooden stopper. The air having been allowed to escape and the iron wire having been drawn back so that it fits snugly against the conical depression in the wood, invert the vessel, bringing it mouth downwards, and hang on the hook K a vessel which can be filled with sand or any heavy material in quantity sufficient to finally separate the upper surface of the stopper, EF, from the lower surface of the water to which it was attached only by the resistance of the vacuum. Next weigh the stopper and wire together with the attached vessel and its contents; we shall then have the force of the vacuum [*forza del vacuo*]. If one attaches to a cylinder of marble

[63]

or glass a weight which, together with the weight of the marble or glass itself, is just equal to the sum of the weights before mentioned, and if breaking occurs we shall then be justified in saying that the vacuum alone holds the parts of the marble and glass together; but if this weight does not suffice and if breaking occurs only after adding, say, four times this weight, we shall then be compelled to say that the vacuum furnishes only one fifth of the total resistance [*resistenza*].

SIMP. No one can doubt the cleverness of the device; yet it presents many difficulties which make me doubt its reliability. For who will assure us that the air does not creep in between the glass and stopper even if it is well packed with tow or other yielding material? I question also whether oiling with wax or turpentine will suffice to make the cone, I, fit snugly on its seat. Besides, may not the parts of the water expand and dilate? Why may not the air or exhalations or some other more subtile substances penetrate the pores of the wood, or even of the glass itself?

SALV. With great skill indeed has Simplicio laid before us the difficulties; and he has even partly suggested how to prevent the
air

air from penetrating the wood or passing between the wood and the glass. But now let me point out that, as our experience increases, we shall learn whether or not these alleged difficulties really exist. For if, as is the case with air, water is by nature expansible, although only under severe treatment, we shall see the stopper descend; and if we put a small excavation in the upper part of the glass vessel, such as indicated by V, then the air or any other tenuous and gaseous substance, which might penetrate the pores of glass or wood, would pass through the water and collect in this receptacle V. But if these things do not happen we may rest assured that our experiment has been performed with proper caution; and we shall discover that water does not dilate and that glass does not allow any material, however tenuous, to penetrate it.

SAGR. Thanks to this discussion, I have learned the cause of a certain effect which I have long wondered at and despaired of understanding. I once saw a cistern which had been provided with a pump under the mistaken impression that the water might thus be drawn with less effort or in greater quantity than by means of the ordinary bucket. The stock of the pump car-

[64]

ried its sucker and valve in the upper part so that the water was lifted by attraction and not by a push as is the case with pumps in which the sucker is placed lower down. This pump worked perfectly so long as the water in the cistern stood above a certain level; but below this level the pump failed to work. When I first noticed this phenomenon I thought the machine was out of order; but the workman whom I called in to repair it told me the defect was not in the pump but in the water which had fallen too low to be raised through such a height; and he added that it was not possible, either by a pump or by any other machine working on the principle of attraction, to lift water a hair's breadth above eighteen cubits; whether the pump be large or small this is the extreme limit of the lift. Up to this time I had been so thoughtless that, although I knew a rope, or rod of wood, or of iron, if sufficiently long, would break by its own weight when held by the upper end, it never occurred to me that

that the same thing would happen, only much more easily, to a column of water. And really is not that thing which is attracted in the pump a column of water attached at the upper end and stretched more and more until finally a point is reached where it breaks, like a rope, on account of its excessive weight?

SALV. That is precisely the way it works; this fixed elevation of eighteen cubits is true for any quantity of water whatever, be the pump large or small or even as fine as a straw. We may therefore say that, on weighing the water contained in a tube eighteen cubits long, no matter what the diameter, we shall obtain the value of the resistance of the vacuum in a cylinder of any solid material having a bore of this same diameter. And having gone so far, let us see how easy it is to find to what length cylinders of metal, stone, wood, glass, etc., of any diameter can be elongated without breaking by their own weight.

[65]

Take for instance a copper wire of any length and thickness; fix the upper end and to the other end attach a greater and greater load until finally the wire breaks; let the maximum load be, say, fifty pounds. Then it is clear that if fifty pounds of copper, in addition to the weight of the wire itself which may be, say, $1/8$ ounce, is drawn out into wire of this same size we shall have the greatest length of this kind of wire which can sustain its own weight. Suppose the wire which breaks to be one cubit in length and $1/8$ ounce in weight; then since it supports 50 lbs. in addition to its own weight, i. e., 4800 eighths-of-an-ounce, it follows that all copper wires, independent of size, can sustain themselves up to a length of 4801 cubits and no more. Since then a copper rod can sustain its own weight up to a length of 4801 cubits it follows that that part of the breaking strength [*resistenza*] which depends upon the vacuum, comparing it with the remaining factors of resistance, is equal to the weight of a rod of water, eighteen cubits long and as thick as the copper rod. If, for example, copper is nine times as heavy as water, the breaking strength [*resistenza allo strapparsi*] of any copper rod, in so far as it depends upon the vacuum, is equal to the weight of two cubits of this same rod. By a similar method one can
find

find the maximum length of wire or rod of any material which will just sustain its own weight, and can at the same time discover the part which the vacuum plays in its breaking strength.

SAGR. It still remains for you to tell us upon what depends the resistance to breaking, other than that of the vacuum; what is the gluey or viscous substance which cements together the parts of the solid? For I cannot imagine a glue that will not burn up in a highly heated furnace in two or three months, or certainly within ten or a hundred. For if gold, silver and glass are kept for a long while in the molten state and are removed from the furnace, their parts, on cooling, immediately reunite and bind themselves together as before. Not only so, but whatever difficulty arises with respect to the cementation of the parts of the glass arises also with regard to the parts of the glue; in other words, what is that which holds these parts together so firmly?

[66]

SALV. A little while ago, I expressed the hope that your good angel might assist you. I now find myself in the same straits. Experiment leaves no doubt that the reason why two plates cannot be separated, except with violent effort, is that they are held together by the resistance of the vacuum; and the same can be said of two large pieces of a marble or bronze column. This being so, I do not see why this same cause may not explain the coherence of smaller parts and indeed of the very smallest particles of these materials. Now, since each effect must have one true and sufficient cause and since I find no other cement, am I not justified in trying to discover whether the vacuum is not a sufficient cause?

SIMP. But seeing that you have already proved that the resistance which the large vacuum offers to the separation of two large parts of a solid is really very small in comparison with that cohesive force which binds together the most minute parts, why do you hesitate to regard this latter as something very different from the former?

SALV. Sagredo has already [p. 13 above] answered this question when he remarked that each individual soldier was being

paid

paid from coin collected by a general tax of pennies and farthings, while even a million of gold would not suffice to pay the entire army. And who knows but that there may be other extremely minute vacua which affect the smallest particles so that that which binds together the contiguous parts is throughout of the same mintage? Let me tell you something which has just occurred to me and which I do not offer as an absolute fact, but rather as a passing thought, still immature and calling for more careful consideration. You may take of it what you like; and judge the rest as you see fit. Sometimes when I have observed how fire winds its way in between the most minute particles of this or that metal and, even though these are solidly cemented together, tears them apart and separates them, and when I have observed that, on removing the fire, these particles reunite with the same tenacity as at first, without any loss of quantity in the case of gold and with little loss in the case of other metals, even though these parts have been separated for a long while, I have thought that the explanation might lie in the fact that the extremely fine particles of fire, penetrating the slender pores of the metal (too small to admit even the finest particles of air or of many other fluids), would fill the small intervening vacua and would set free these small particles from the attraction which these same vacua exert upon them and which prevents their separation. Thus the particles are able to

[67]

move freely so that the mass [*massa*] becomes fluid and remains so as long as the particles of fire remain inside; but if they depart and leave the former vacua then the original attraction [*attraxzione*] returns and the parts are again cemented together.

In reply to the question raised by Simplicio, one may say that although each particular vacuum is exceedingly minute and therefore easily overcome, yet their number is so extraordinarily great that their combined resistance is, so to speak, multiplied almost without limit. The nature and the amount of force [*forza*] which results [*risulta*] from adding together an immense number of small forces [*debolissimi momenti*] is clearly illustrated by the fact that a weight of millions of pounds, suspended by by great cables, is overcome and lifted, when the south wind carries innumerable atoms of water, suspended in thin mist, which moving through the air penetrate between the fibres of the tense ropes in spite of the tremendous force of the hanging weight. When these particles enter the narrow pores they swell the ropes, thereby shorten them, and perforce lift the heavy mass [mole].

SAGR. There can be no doubt that any resistance, so long as it is not infinite, may be overcome by a multitude of minute forces. Thus a vast number of ants might carry ashore a ship laden with grain. And since experience shows us daily that one ant can easily carry one grain, it is clear that the number of grains in the ship is not infinite, but falls below a certain limit. If you take another number four or six times as great, and if you set to work a corresponding number of ants they will carry the grain ashore and the boat also. It is true that this will call for a prodigious number of ants, but in my opinion this is precisely the case with the vacua which bind together the least particles of a metal.

SALV. But even if this demanded an infinite number would you still think it impossible?

SAGR. Not if the mass [mole] of metal were infinite; otherwise. . . .

[68]

SALV. Otherwise what? Now since we have arrived at paradoxes let us see if we cannot prove that within a finite extent it is possible to discover an infinite number of vacua. At the same time we shall at least reach a solution of the most remarkable of all that list of problems which Aristotle himself calls wonderful; I refer to his *Questions in Mechanics*. This solution may be no less clear and conclusive than that which he himself gives and quite different also from that so cleverly expounded by the most learned Monsignor di Guevara.*

First it is necessary to consider a proposition, not treated by others, but upon which depends the solution of the problem and from which, if I mistake not, we shall derive other new and remarkable facts. For the sake of clearness let us draw an

* Bishop of Teano; b. 1561, d.1641. [*Trans.*]

accurate figure. About G as a center describe an equiangular and equilateral polygon of any number of sides, say the hexagon ABCDEF. Similar to this and concentric with it, describe another smaller one which we shall call HIKLMN. Prolong the

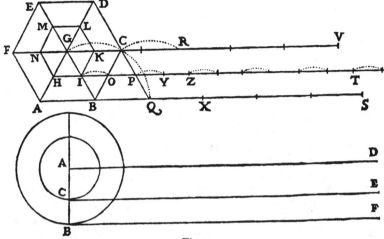

Fig. 5

side AB, of the larger hexagon, indefinitely toward S; in like manner prolong the corresponding side HI of the smaller hexagon, in the same direction, so that the line HT is parallel to AS; and through the center draw the line GV parallel to the other two. This done, imagine the larger polygon to roll upon
[69]
the line AS, carrying with it the smaller polygon. It is evident that, if the point B, the end of the side AB, remains fixed at the beginning of the rotation, the point A will rise and the point C will fall describing the arc CQ until the side BC coincides with the line BQ, equal to BC. But during this rotation the point I, on the smaller polygon, will rise above the line IT because IB is oblique to AS; and it will not again return to the line IT until the point C shall have reached the position Q. The point I, having described the arc IO above the line HT, will reach the position
O at

O at the same time the side IK assumes the position OP; but in the meantime the center G has traversed a path above GV and does not return to it until it has completed the arc GC. This step having been taken, the larger polygon has been brought to rest with its side BC coinciding with the line BQ while the side IK of the smaller polygon has been made to coincide with the line OP, having passed over the portion IO without touching it; also the center G will have reached the position C after having traversed all its course above the parallel line GV. And finally the entire figure will assume a position similar to the first, so that if we continue the rotation and come to the next step, the side DC of the larger polygon will coincide with the portion QX and the side KL of the smaller polygon, having first skipped the arc PY, will fall on YZ, while the center still keeping above the line GV will return to it at R after having jumped the interval CR. At the end of one complete rotation the larger polygon will have traced upon the line AS, without break, six lines together equal to its perimeter; the lesser polygon will likewise have imprinted six lines equal to its perimeter, but separated by the interposition of five arcs, whose chords represent the parts of HT not touched by the polygon: the center G never reaches the line GV except at six points. From this it is clear that the space traversed by the smaller polygon is almost equal to that traversed by the larger, that is, the line HT approximates the line AS, differing from it only by the length of one chord of one of these arcs, provided we understand the line HT to include the five skipped arcs.

Now this exposition which I have given in the case of these hexagons must be understood to be applicable to all other polygons, whatever the number of sides, provided only they are

[70]

similar, concentric, and rigidly connected, so that when the greater one rotates the lesser will also turn however small it may be. You must also understand that the lines described by these two are nearly equal provided we include in the space traversed by the smaller one the intervals which are not touched by any part of the perimeter of this smaller polygon.

Let

Let a large polygon of, say, one thousand sides make one
complete rotation and thus lay off a line equal to its perimeter;
at the same time the small one will pass over an approximately
equal distance, made up of a thousand small portions each
equal to one of its sides, but interrupted by a thousand spaces
which, in contrast with the portions that coincide with the sides
of the polygon, we may call empty. So far the matter is free
from difficulty or doubt.

But now suppose that about any center, say A, we describe
two concentric and rigidly connected circles; and suppose that
from the points C and B, on their radii, there are drawn the
tangents CE and BF and that through the center A the line AD
is drawn parallel to them, then if the large circle makes one
complete rotation along the line BF, equal not only to its circumference but also to the other two lines CE and AD, tell me
what the smaller circle will do and also what the center will do.
As to the center it will certainly traverse and touch the entire
line AD while the circumference of the smaller circle will have
measured off by its points of contact the entire line CE, just as
was done by the above mentioned polygons. The only difference
is that the line HT was not at every point in contact with the
perimeter of the smaller polygon, but there were left untouched
as many vacant spaces as there were spaces coinciding with the
sides. But here in the case of the circles the circumference of the
smaller one never leaves the line CE, so that no part of the latter
is left untouched, nor is there ever a time when some point on the
circle is not in contact with the straight line. How now can the
smaller circle traverse a length greater than its circumference
unless it go by jumps?

SAGR. It seems to me that one may say that just as the center
of the circle, by itself, carried along the line AD is constantly in
contact with it, although it is only a single point, so the points on
the circumference of the smaller circle, carried along by the
motion of the larger circle, would slide over some small parts of
the line CE.

[71]

SALV. There are two reasons why this cannot happen. First
because

because there is no ground for thinking that one point of contact, such as that at C, rather than another, should slip over
certain portions of the line CE. But if such slidings along CE
did occur they would be infinite in number since the points of
contact (being mere points) are infinite in number: an infinite
number of finite slips will however make an infinitely long line,
while as a matter of fact the line CE is finite. The other reason
is that as the greater circle, in its rotation, changes its point of
contact continuously the lesser circle must do the same because
B is the only point from which a straight line can be drawn to A
and pass through C. Accordingly the small circle must change
its point of contact whenever the large one changes: no point of
the small circle touches the straight line CE in more than one
point. Not only so, but even in the rotation of the polygons
there was no point on the perimeter of the smaller which coincided with more than one point on the line traversed by that
perimeter; this is at once clear when you remember that the
line IK is parallel to BC and that therefore IK will remain above
IP until BC coincides with BQ, and that IK will not lie upon IP
except at the very instant when BC occupies the position BQ; at
this instant the entire line IK coincides with OP and immediately
afterwards rises above it.

SAGR. This is a very intricate matter. I see no solution. Pray
explain it to us.

SALV. Let us return to the consideration of the above mentioned polygons whose behavior we already understand. Now
in the case of polygons with 100000 sides, the line traversed by
the perimeter of the greater, i. e., the line laid down by its
100000 sides one after another, is equal to the line traced out by
the 100000 sides of the smaller, provided we include the 100000
vacant spaces interspersed. So in the case of the circles, polygons having an infinitude of sides, the line traversed by the
continuously distributed [*continuamente disposti*] infinitude of
sides is in the greater circle equal to the line laid down by the
infinitude of sides in the smaller circle but with the exception
that these latter alternate with empty spaces; and since the
sides are not finite in number, but infinite, so also are the intervening

vening empty spaces not finite but infinite. The line traversed
by the larger circle consists then of an infinite number of points
which completely fill it; while that which is traced by the smaller
circle consists of an infinite number of points which leave empty
spaces and only partly fill the line. And here I wish you to
observe that after dividing and resolving a line into a finite
number of parts, that is, into a number which can be counted, it

[72]

is not possible to arrange them again into a greater length than
that which they occupied when they formed a *continuum* [*continuate*] and were connected without the interposition of as
many empty spaces. But if we consider the line resolved into
an infinite number of infinitely small and indivisible parts, we
shall be able to conceive the line extended indefinitely by the
interposition, not of a finite, but of an infinite number of infinitely small indivisible empty spaces.

Now this which has been said concerning simple lines must be
understood to hold also in the case of surfaces and solid bodies,
it being assumed that they are made up of an infinite, not a
finite, number of atoms. Such a body once divided into a
finite number of parts it is impossible to reassemble them so as to
occupy more space than before unless we interpose a finite
number of empty spaces, that is to say, spaces free from the
substance of which the solid is made. But if we imagine the
body, by some extreme and final analysis, resolved into its
primary elements, infinite in number, then we shall be able to
think of them as indefinitely extended in space, not by the
interposition of a finite, but of an infinite number of empty
spaces. Thus one can easily imagine a small ball of gold expanded into a very large space without the introduction of a
finite number of empty spaces, always provided the gold is
made up of an infinite number of indivisible parts.

SIMP. It seems to me that you are travelling along toward
those vacua advocated by a certain ancient philosopher.

SALV. But you have failed to add, "who denied Divine Providence," an inapt remark made on a similar occasion by a certain antagonist of our Academician.

Simp.

SIMP. I noticed, and not without indignation, the rancor of this ill-natured opponent; further references to these affairs I omit, not only as a matter of good form, but also because I know how unpleasant they are to the good tempered and well ordered mind of one so religious and pious, so orthodox and God-fearing as you.

But to return to our subject, your previous discourse leaves with me many difficulties which I am unable to solve. First among these is that, if the circumferences of the two circles are equal to the two straight lines, CE and BF, the latter considered as a *continuum*, the former as interrupted with an infinity of empty points, I do not see how it is possible to say that the line AD described by the center, and made up of an infinity of points, is equal to this center which is a single point. Besides, this building up of lines out of points, divisibles out of indivisibles, and finites out of infinites, offers me an obstacle difficult to avoid; and the necessity of introducing a vacuum, so conclusively refuted by Aristotle, presents the same difficulty.

[73]

SALV. These difficulties are real; and they are not the only ones. But let us remember that we are dealing with infinities and indivisibles, both of which transcend our finite understanding, the former on account of their magnitude, the latter because of their smallness. In spite of this, men cannot refrain from discussing them, even though it must be done in a roundabout way.

Therefore I also should like to take the liberty to present some of my ideas which, though not necessarily convincing, would, on account of their novelty, at least, prove somewhat startling. But such a diversion might perhaps carry us too far away from the subject under discussion and might therefore appear to you inopportune and not very pleasing.

SAGR. Pray let us enjoy the advantages and privileges which come from conversation between friends, especially upon subjects freely chosen and not forced upon us, a matter vastly different from dealing with dead books which give rise to many doubts but remove none. Share with us, therefore, the thoughts which

which our discussion has suggested to you; for since we are free from urgent business there will be abundant time to pursue the topics already mentioned; and in particular the objections raised by Simplicio ought not in any wise to be neglected.

SALV. Granted, since you so desire. The first question was, How can a single point be equal to a line? Since I cannot do more at present I shall attempt to remove, or at least diminish, one improbability by introducing a similar or a greater one, just as sometimes a wonder is diminished by a miracle.*

And this I shall do by showing you two equal surfaces, together with two equal solids located upon these same surfaces as bases, all four of which diminish continuously and uniformly in such a way that their remainders always preserve equality among themselves, and finally both the surfaces and the solids terminate their previous constant equality by degenerating, the one solid and the one surface into a very long line, the other solid and the other surface into a single point; that is, the latter to one point, the former to an infinite number of points.

[74]

SAGR. This proposition appears to me wonderful, indeed; but let us hear the explanation and demonstration.

SALV. Since the proof is purely geometrical we shall need a figure. Let AFB be a semicircle with center at C; about it describe the rectangle ADEB and from the center draw the straight lines CD and CE to the points D and E. Imagine the radius CF to be drawn perpendicular to either of the lines AB or DE, and the entire figure to rotate about this radius as an axis. It is clear that the rectangle ADEB will thus describe a cylinder, the semicircle AFB a hemisphere, and the triangle CDE, a cone. Next let us remove the hemisphere but leave the cone and the rest of the cylinder, which, on account of its shape, we will call a "bowl." First we shall prove that the bowl and the cone are equal; then we shall show that a plane drawn parallel to the circle which forms the base of the bowl and which has the line DE for diameter and F for a center—a plane whose trace is GN—cuts the bowl in the points G, I, O, N, and the cone in the points H, L, so that the part of the cone indicated by CHL is always equal to

* Cf. p. 30 below. [Trans.]

the part of the bowl whose profile is represented by the triangles GAI and BON. Besides this we shall prove that the base of the cone, i. e., the circle whose diameter is HL, is equal to the circular

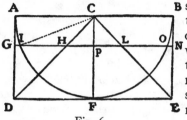 surface which forms the base of this portion of the bowl, or as one might say, equal to a ribbon whose width is GI. (Note by the way the nature of mathematical definitions which consist merely in the imposition of names or, if you prefer, abbreviations of speech established and

Fig. 6

introduced in order to avoid the tedious drudgery which you and I now experience simply because we have not agreed to call this surface a "circular band" and that sharp solid portion of the bowl a "round razor.") Now call them by

[75]

what name you please, it suffices to understand that the plane, drawn at any height whatever, so long as it is parallel to the base, i. e., to the circle whose diameter is DE, always cuts the two solids so that the portion CHL of the cone is equal to the upper portion of the bowl; likewise the two areas which are the bases of these solids, namely the band and the circle HL, are also equal. Here we have the miracle mentioned above; as the cutting plane approaches the line AB the portions of the solids cut off are always equal, so also the areas of their bases. And as the cutting plane comes near the top, the two solids (always equal) as well as their bases (areas which are also equal) finally vanish, one pair of them degenerating into the circumference of a circle, the other into a single point, namely, the upper edge of the bowl and the apex of the cone. Now, since as these solids diminish equality is maintained between them up to the very last, we are justified in saying that, at the extreme and final end of this diminution, they are still equal and that one is not infinitely greater than the other. It appears therefore that we may equate the circumference of a large circle to a single point. And this which is true of the solids is true also of the surfaces which

form

form their bases; for these also preserve equality between themselves throughout their diminution and in the end vanish, the one into the circumference of a circle, the other into a single point. Shall we not then call them equal seeing that they are the last traces and remnants of equal magnitudes? Note also that, even if these vessels were large enough to contain immense celestial hemispheres, both their upper edges and the apexes of the cones therein contained would always remain equal and would vanish, the former into circles having the dimensions of the largest celestial orbits, the latter into single points. Hence in conformity with the preceding we may say that all circumferences of circles, however different, are equal to each other, and are each equal to a single point.

SAGR. This presentation strikes me as so clever and novel that, even if I were able, I would not be willing to oppose it; for to deface so beautiful a structure by a blunt pedantic attack would be nothing short of sinful. But for our complete satisfac-

[76]

tion pray give us this geometrical proof that there is always equality between these solids and between their bases; for it cannot, I think, fail to be very ingenious, seeing how subtle is the philosophical argument based upon this result.

SALV. The demonstration is both short and easy. Referring to the preceding figure, since IPC is a right angle the square of the radius IC is equal to the sum of the squares on the two sides IP, PC; but the radius IC is equal to AC and also to GP, while CP is equal to PH. Hence the square of the line GP is equal to the sum of the squares of IP and PH, or multiplying through by 4, we have the square of the diameter GN equal to the sum of the squares on IO and HL. And, since the areas of circles are to each other as the squares of their diameters, it follows that the area of the circle whose diameter is GN is equal to the sum of the areas of circles having diameters IO and HL, so that if we remove the common area of the circle having IO for diameter the remaining area of the circle GN will be equal to the area of the circle whose diameter is HL. So much for the first part. As for the other part, we leave its demonstration for the present, partly because

because those who wish to follow it will find it in the twelfth proposition of the second book of *De centro gravitatis solidorum* by the Archimedes of our age, Luca Valerio,* who made use of it for a different object, and partly because, for our purpose, it suffices to have seen that the above-mentioned surfaces are always equal and that, as they keep on diminishing uniformly, they degenerate, the one into a single point, the other into the circumference of a circle larger than any assignable; in this fact lies our miracle.†

SAGR. The demonstration is ingenious and the inferences drawn from it are remarkable. And now let us hear something concerning the other difficulty raised by Simplicio, if you have anything special to say, which, however, seems to me hardly possible, since the matter has already been so thoroughly discussed.

SALV. But I do have something special to say, and will first of all repeat what I said a little while ago, namely, that infinity and indivisibility are in their very nature incomprehensible to us; imagine then what they are when combined. Yet if

[77]

we wish to build up a line out of indivisible points, we must take an infinite number of them, and are, therefore, bound to understand both the infinite and the indivisible at the same time. Many ideas have passed through my mind concerning this subject, some of which, possibly the more important, I may not be able to recall on the spur of the moment; but in the course of our discussion it may happen that I shall awaken in you, and especially in Simplicio, objections and difficulties which in turn will bring to memory that which, without such stimulus, would have lain dormant in my mind. Allow me therefore the customary liberty of introducing some of our human fancies, for indeed we may so call them in comparison with supernatural truth which furnishes the one true and safe recourse for decision in our discussions and which is an infallible guide in the dark and dubious paths of thought.

* Distinguished Italian mathematician; born at Ferrara about 1552; admitted to the Accademia dei Lincei 1612; died 1618. [*Trans.*]

† *Cf.* p. 27 above. [*Trans.*]

One of the main objections urged against this building up of continuous quantities out of indivisible quantities [*continuo d' indivisibili*] is that the addition of one indivisible to another cannot produce a divisible, for if this were so it would render the indivisible divisible. Thus if two indivisibles, say two points, can be united to form a quantity, say a divisible line, then an even more divisible line might be formed by the union of three, five, seven, or any other odd number of points. Since however these lines can be cut into two equal parts, it becomes possible to cut the indivisible which lies exactly in the middle of the line. In answer to this and other objections of the same type we reply that a divisible magnitude cannot be constructed out of two or ten or a hundred or a thousand indivisibles, but requires an infinite number of them.

SIMP. Here a difficulty presents itself which appears to me insoluble. Since it is clear that we may have one line greater than another, each containing an infinite number of points, we are forced to admit that, within one and the same class, we may have something greater than infinity, because the infinity of points in the long line is greater than the infinity of points in the short line. This assigning to an infinite quantity a value greater than infinity is quite beyond my comprehension.

SALV. This is one of the difficulties which arise when we attempt, with our finite minds, to discuss the infinite, assigning to it those properties which we give to the finite and limited; but
[78]
this I think is wrong, for we cannot speak of infinite quantities as being the one greater or less than or equal to another. To prove this I have in mind an argument which, for the sake of clearness, I shall put in the form of questions to Simplicio who raised this difficulty.

I take it for granted that you know which of the numbers are squares and which are not.

SIMP. I am quite aware that a squared number is one which results from the multiplication of another number by itself; thus 4, 9, etc., are squared numbers which come from multiplying 2, 3, etc., by themselves.

Salv.

SALV. Very well; and you also know that just as the products are called squares so the factors are called sides or roots; while on the other hand those numbers which do not consist of two equal factors are not squares. Therefore if I assert that all numbers, including both squares and non-squares, are more than the squares alone, I shall speak the truth, shall I not?

SIMP. Most certainly.

SALV. If I should ask further how many squares there are one might reply truly that there are as many as the corresponding number of roots, since every square has its own root and every root its own square, while no square has more than one root and no root more than one square.

SIMP. Precisely so.

SALV. But if I inquire how many roots there are, it cannot be denied that there are as many as there are numbers because every number is a root of some square. This being granted we must say that there are as many squares as there are numbers because they are just as numerous as their roots, and all the numbers are roots. Yet at the outset we said there are many more numbers than squares, since the larger portion of them are not squares. Not only so, but the proportionate number of squares diminishes as we pass to larger numbers. Thus up to 100 we have 10 squares, that is, the squares constitute 1/10 part of all the numbers; up to 10000, we find only 1/100

[79]

part to be squares; and up to a million only 1/1000 part; on the other hand in an infinite number, if one could conceive of such a thing, he would be forced to admit that there are as many squares as there are numbers all taken together.

SAGR. What then must one conclude under these circumstances?

SALV. So far as I see we can only infer that the totality of all numbers is infinite, that the number of squares is infinite, and that the number of their roots is infinite; neither is the number of squares less than the totality of all numbers, nor the latter greater than the former; and finally the attributes "equal," "greater," and "less," are not applicable to infinite, but

but only to finite, quantities. When therefore Simplicio introduces several lines of different lengths and asks me how it is possible that the longer ones do not contain more points than the shorter, I answer him that one line does not contain more or less or just as many points as another, but that each line contains an infinite number. Or if I had replied to him that the points in one line were equal in number to the squares; in another, greater than the totality of numbers; and in the little one, as many as the number of cubes, might I not, indeed, have satisfied him by thus placing more points in one line than in another and yet maintaining an infinite number in each? So much for the first difficulty.

SAGR. Pray stop a moment and let me add to what has already been said an idea which just occurs to me. If the preceding be true, it seems to me impossible to say either that one infinite number is greater than another or even that it is greater than a finite number, because if the infinite number were greater than, say, a million it would follow that on passing from the million to higher and higher numbers we would be approaching the infinite; but this is not so; on the contrary, the larger the number to which we pass, the more we recede from [this property of] infinity, because the greater the numbers the fewer [relatively] are the squares contained in them; but the squares in infinity cannot be less than the totality of all the numbers, as we have just agreed; hence the approach to greater and greater numbers means a departure from infinity.*

SALV. And thus from your ingenious argument we are led to

[80]

conclude that the attributes "larger," "smaller," and "equal" have no place either in comparing infinite quantities with each other or in comparing infinite with finite quantities.

I pass now to another consideration. Since lines and all continuous quantities are divisible into parts which are themselves divisible without end, I do not see how it is possible

---

* A certain confusion of thought appears to be introduced here through a failure to distinguish between the *number n* and the *class* of the first *n* numbers; and likewise from a failure to distinguish infinity as a number from infinity as the class of all numbers. [*Trans.*]

to avoid the conclusion that these lines are built up of an infinite number of indivisible quantities because a division and a subdivision which can be carried on indefinitely presupposes that the parts are infinite in number, otherwise the subdivision would reach an end; and if the parts are infinite in number, we must conclude that they are not finite in size, because an infinite number of finite quantities would give an infinite magnitude. And thus we have a continuous quantity built up of an infinite number of indivisibles.

SIMP. But if we can carry on indefinitely the division into finite parts what necessity is there then for the introduction of non-finite parts?

SALV. The very fact that one is able to continue, without end, the division into finite parts [*in parti quante*] makes it necessary to regard the quantity as composed of an infinite number of immeasurably small elements [*di infiniti non quanti*]. Now in order to settle this matter I shall ask you to tell me whether, in your opinion, a *continuum* is made up of a finite or of an infinite number of finite parts [*parti quante*].

SIMP. My answer is that their number is both infinite and finite; potentially infinite but actually finite [*infinite, in potenza; e finite, in atto*]; that is to say, potentially infinite before division and actually finite after division; because parts cannot be said to exist in a body which is not yet divided or at least marked out; if this is not done we say that they exist potentially.

SALV. So that a line which is, for instance, twenty spans long is not said to contain actually twenty lines each one span in length except after division into twenty equal parts; before division it is said to contain them only potentially. Suppose the facts are as you say; tell me then whether, when the division is once made, the size of the original quantity is thereby increased, diminished, or unaffected.

SIMP. It neither increases nor diminishes.

SALV. That is my opinion also. Therefore the finite parts [*parti quante*] in a *continuum*, whether actually or potentially present, do not make the quantity either larger or smaller; but it is perfectly clear that, if the number of finite parts actually
contained

contained in the whole is infinite in number, they will make the magnitude infinite. Hence the number of finite parts, although existing only potentially, cannot be infinite unless the magnitude containing them be infinite; and conversely if the magnitude is

[81]

finite it cannot contain an infinite number of finite parts either actually or potentially.

SAGR. How then is it possible to divide a *continuum* without limit into parts which are themselves always capable of subdivision?

SALV. This distinction of yours between actual and potential appears to render easy by one method what would be impossible by another. But I shall endeavor to reconcile these matters in another way; and as to the query whether the finite parts of a limited *continuum* [*continuo terminato*] are finite or infinite in number I will, contrary to the opinion of Simplicio, answer that they are neither finite nor infinite.

SIMP. This answer would never have occurred to me since I did not think that there existed any intermediate step between the finite and the infinite, so that the classification or distinction which assumes that a thing must be either finite or infinite is faulty and defective.

SALV. So it seems to me. And if we consider discrete quantities I think there is, between finite and infinite quantities, a third intermediate term which corresponds to every assigned number; so that if asked, as in the present case, whether the finite parts of a *continuum* are finite or infinite in number the best reply is that they are neither finite nor infinite but correspond to every assigned number. In order that this may be possible, it is necessary that those parts should not be included within a limited number, for in that case they would not correspond to a number which is greater; nor can they be infinite in number since no assigned number is infinite; and thus at the pleasure of the questioner we may, to any given line, assign a hundred finite parts, a thousand, a hundred thousand, or indeed any number we may please so long as it be not infinite. I grant, therefore, to the philosophers, that the *continuum* contains as

many

many finite parts as they please and I concede also that it contains them, either actually or potentially, as they may like; but I must add that just as a line ten fathoms [*canne*] in length contains ten lines each of one fathom and forty lines each of one cubit [*braccia*] and eighty lines each of half a cubit, etc., so it contains an infinite number of points; call them actual or potential, as you like, for as to this detail, Simplicio, I defer to your opinion and to your judgment.

[82]

Simp. I cannot help admiring your discussion; but I fear that this parallelism between the points and the finite parts contained in a line will not prove satisfactory, and that you will not find it so easy to divide a given line into an infinite number of points as the philosophers do to cut it into ten fathoms or forty cubits; not only so, but such a division is quite impossible to realize in practice, so that this will be one of those potentialities which cannot be reduced to actuality.

Salv. The fact that something can be done only with effort or diligence or with great expenditure of time does not render it impossible; for I think that you yourself could not easily divide a line into a thousand parts, and much less if the number of parts were 937 or any other large prime number. But if I were to accomplish this division which you deem impossible as readily as another person would divide the line into forty parts would you then be more willing, in our discussion, to concede the possibility of such a division?

Simp. In general I enjoy greatly your method; and replying to your query, I answer that it would be more than sufficient if it prove not more difficult to resolve a line into points than to divide it into a thousand parts.

Salv. I will now say something which may perhaps astonish you; it refers to the possibility of dividing a line into its infinitely small elements by following the same order which one employs in dividing the same line into forty, sixty, or a hundred parts, that is, by dividing it into two, four, etc. He who thinks that, by following this method, he can reach an infinite number of points is greatly mistaken; for if this process were followed to eternity

eternity there would still remain finite parts which were undivided.

Indeed by such a method one is very far from reaching the goal of indivisibility; on the contrary he recedes from it and while he thinks that, by continuing this division and by multiplying the multitude of parts, he will approach infinity, he is, in my opinion, getting farther and farther away from it. My reason is this. In the preceding discussion we concluded that, in an infinite number, it is necessary that the squares and cubes should be as numerous as the totality of the natural numbers [*tutti i numeri*], because both of these are as numerous as their roots which constitute the totality of the natural numbers. Next we saw that the larger the numbers taken the more sparsely distributed were the squares, and still more sparsely the cubes; therefore it is clear that the larger the numbers to which we pass the farther we recede from the infinite number; hence it follows

[83]

that, since this process carries us farther and farther from the end sought, if on turning back we shall find that any number can be said to be infinite, it must be unity. Here indeed are satisfied all those conditions which are requisite for an infinite number; I mean that unity contains in itself as many squares as there are cubes and natural numbers [*tutti i numeri*].

SIMP. I do not quite grasp the meaning of this.

SALV. There is no difficulty in the matter because unity is at once a square, a cube, a square of a square and all the other powers [*dignità*]; nor is there any essential peculiarity in squares or cubes which does not belong to unity; as, for example, the property of two square numbers that they have between them a mean proportional; take any square number you please as the first term and unity for the other, then you will always find a number which is a mean proportional. Consider the two square numbers, 9 and 4; then 3 is the mean proportional between 9 and 1; while 2 is a mean proportional between 4 and 1; between 9 and 4 we have 6 as a mean proportional. A property of cubes is that they must have between them two mean proportional numbers; take 8 and 27; between them lie 12 and 18; while between

between 1 and 8 we have 2 and 4 intervening; and between 1 and 27 there lie 3 and 9. Therefore we conclude that unity is the only infinite number. These are some of the marvels which our imagination cannot grasp and which should warn us against the serious error of those who attempt to discuss the infinite by assigning to it the same properties which we employ for the finite, the natures of the two having nothing in common.

With regard to this subject I must tell you of a remarkable property which just now occurs to me and which will explain the vast alteration and change of character which a finite quantity would undergo in passing to infinity. Let us draw the straight line AB of arbitrary length and let the point C divide it into two unequal parts; then I say that, if pairs of lines be drawn, one from each of the terminal points A and B, and if the ratio between the lengths of these lines is the same as that between AC and CB, their points of intersection will all lie upon the circumference of one and the same circle. Thus, for ex-

[84]

ample, AL and BL drawn from A and B, meeting at the point L, bearing to one another the same ratio as AC to BC, and the

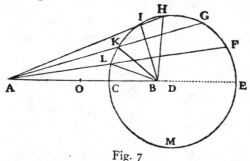

Fig. 7

pair AK and BK meeting at K also bearing to one another the same ratio, and likewise the pairs AI, BI, AH, BH, AG, BG, AF, BF, AE, BE, have their points of intersection L, K, I, H, G, F, E, all lying upon the circumference of one and the same circle. Accordingly if we imagine the point C to move continuously in such a manner that the lines drawn from it to the fixed terminal points, A and B, always maintain the same ratio between their lengths as exists between the original parts, AC and CB, then the point C will, as I shall presently prove, describe a circle. And the circle thus described will

increase

increase in size without limit as the point C approaches the middle point which we may call O; but it will diminish in size as C approaches the end B. So that the infinite number of points located in the line OB will, if the motion be as explained above, describe circles of every size, some smaller than the pupil of the eye of a flea, others larger than the celestial equator. Now if we move any of the points lying between the two ends O and B they will all describe circles, those nearest O, immense circles; but if we move the point O itself, and continue to move it according to the aforesaid law, namely, that the lines drawn from O to the terminal points, A and B, maintain the same ratio as the original lines AO and OB, what kind of a line will be produced? A circle will be drawn larger than the largest of the others, a circle which is therefore infinite. But from the point O a straight line will also be drawn perpendicular to BA and extending to infinity without ever turning, as did the others, to join its last end with its first; for the point C, with its limited motion, having described

[85]

the upper semi-circle, CHE, proceeds to describe the lower semicircle EMC, thus returning to the starting point. But the point O having started to describe its circle, as did all the other points in the line AB, (for the points in the other portion OA describe their circles also, the largest being those nearest the point O) is unable to return to its starting point because the circle it describes, being the largest of all, is infinite; in fact, it describes an infinite straight line as circumference of its infinite circle. Think now what a difference there is between a finite and an infinite circle since the latter changes character in such a manner that it loses not only its existence but also its possibility of existence; indeed, we already clearly understand that there can be no such thing as an infinite circle; similarly there can be no infinite sphere, no infinite body, and no infinite surface of any shape. Now what shall we say concerning this metamorphosis in the transition from finite to infinite? And why should we feel greater repugnance, seeing that, in our search after the infinite among numbers we found it in unity? Having broken up a solid into many parts, having reduced it to the finest of

powder

powder and having resolved it into its infinitely small indivisible atoms why may we not say that this solid has been reduced to a single *continuum* [*un solo continuo*] perhaps a fluid like water or mercury or even a liquified metal? And do we not see stones melt into glass and the glass itself under strong heat become more fluid than water?

SAGR. Are we then to believe that substances become fluid in virtue of being resolved into their infinitely small indivisible components?

SALV. I am not able to find any better means of accounting for certain phenomena of which the following is one. When I take a hard substance such as stone or metal and when I reduce it by means of a hammer or fine file to the most minute and impalpable powder, it is clear that its finest particles, although when taken one by one are, on account of their smallness, imperceptible to our sight and touch, are nevertheless finite in size, possess shape, and capability of being counted. It is also true that when once heaped up they remain in a heap; and if an excavation be made within limits the cavity will remain and the surrounding particles will not rush in to fill it; if shaken the particles come to rest immediately after the external disturbing agent is removed; the same effects are observed in all piles of

[86]

larger and larger particles, of any shape, even if spherical, as is the case with piles of millet, wheat, lead shot, and every other material. But if we attempt to discover such properties in water we do not find them; for when once heaped up it immediately flattens out unless held up by some vessel or other external retaining body; when hollowed out it quickly rushes in to fill the cavity; and when disturbed it fluctuates for a long time and sends out its waves through great distances.

Seeing that water has less firmness [*consistenza*] than the finest of powder, in fact has no consistence whatever, we may, it seems to me, very reasonably conclude that the smallest particles into which it can be resolved are quite different from finite and divisible particles; indeed the only difference I am able to discover is that the former are indivisible. The exquisite transparency

transparency of water also favors this view; for the most transparent crystal when broken and ground and reduced to powder loses its transparency; the finer the grinding the greater the loss; but in the case of water where the attrition is of the highest degree we have extreme transparency. Gold and silver when pulverized with acids [*acque forti*] more finely than is possible with any file still remain powders,* and do not become fluids until the finest particles [*gl' indivisibili*] of fire or of the rays of the sun dissolve them, as I think, into their ultimate, indivisible, and infinitely small components.

SAGR. This phenomenon of light which you mention is one which I have many times remarked with astonishment. I have, for instance, seen lead melted instantly by means of a concave mirror only three hands [*palmi*] in diameter. Hence I think that if the mirror were very large, well-polished and of a parabolic figure, it would just as readily and quickly melt any other metal, seeing that the small mirror, which was not well polished and had only a spherical shape, was able so energetically to melt lead and burn every combustible substance. Such effects as these render credible to me the marvels accomplished by the mirrors of Archimedes.

SALV. Speaking of the effects produced by the mirrors of Archimedes, it was his own books (which I had already read and studied with infinite astonishment) that rendered credible to me all the miracles described by various writers. And if any doubt had remained the book which Father Buonaventura Cavalieri†

[87]

has recently published on the subject of the burning glass [*specchio ustorio*] and which I have read with admiration would have removed the last difficulty.

SAGR. I also have seen this treatise and have read it with

* It is not clear what Galileo here means by saying that gold and silver when treated with acids still remain powders. [*Trans.*]

† One of the most active investigators among Galileo's contemporaries; born at Milan 1598; died at Bologna 1647; a Jesuit father, first to introduce the use of logarithms into Italy and first to derive the expression for the focal length of a lens having unequal radii of curvature. His "method of indivisibles" is to be reckoned as a precursor of the infinitesimal calculus. [*Trans.*]

pleasure and astonishment; and knowing the author I was confirmed in the opinion which I had already formed of him that he was destined to become one of the leading mathematicians of our age. But now, with regard to the surprising effect of solar rays in melting metals, must we believe that such a furious action is devoid of motion or that it is accompanied by the most rapid of motions?

SALV. We observe that other combustions and resolutions are accompanied by motion, and that, the most rapid; note the action of lightning and of powder as used in mines and petards; note also how the charcoal flame, mixed as it is with heavy and impure vapors, increases its power to liquify metals whenever quickened by a pair of bellows. Hence I do not understand how the action of light, although very pure, can be devoid of motion and that of the swiftest type.

SAGR. But of what kind and how great must we consider this speed of light to be? Is it instantaneous or momentary or does it like other motions require time? Can we not decide this by experiment?

SIMP. Everyday experience shows that the propagation of light is instantaneous; for when we see a piece of artillery fired, at great distance, the flash reaches our eyes without lapse of time; but the sound reaches the ear only after a noticeable interval.

SAGR. Well, Simplicio, the only thing I am able to infer from this familiar bit of experience is that sound, in reaching our ear, travels more slowly than light; it does not inform me whether the coming of the light is instantaneous or whether, although extremely rapid, it still occupies time. An observation of this kind tells us nothing more than one in which it is claimed that "As soon as the sun reaches the horizon its light reaches our eyes"; but who will assure me that these rays had not reached this limit earlier than they reached our vision?

SALV. The small conclusiveness of these and other similar observations once led me to devise a method by which one might accurately ascertain whether illumination, i. e., the propagation of light, is really instantaneous. The fact that the speed of sound

[88]

sound is as high as it is, assures us that the motion of light cannot fail to be extraordinarily swift. The experiment which I devised was as follows:

Let each of two persons take a light contained in a lantern, or other receptacle, such that by the interposition of the hand, the one can shut off or admit the light to the vision of the other. Next let them stand opposite each other at a distance of a few cubits and practice until they acquire such skill in uncovering and occulting their lights that the instant one sees the light of his companion he will uncover his own. After a few trials the response will be so prompt that without sensible error [*svario*] the uncovering of one light is immediately followed by the uncovering of the other, so that as soon as one exposes his light he will instantly see that of the other. Having acquired skill at this short distance let the two experimenters, equipped as before, take up positions separated by a distance of two or three miles and let them perform the same experiment at night, noting carefully whether the exposures and occultations occur in the same manner as at short distances; if they do, we may safely conclude that the propagation of light is instantaneous; but if time is required at a distance of three miles which, considering the going of one light and the coming of the other, really amounts to six, then the delay ought to be easily observable. If the experiment is to be made at still greater distances, say eight or ten miles, telescopes may be employed, each observer adjusting one for himself at the place where he is to make the experiment at night; then although the lights are not large and are therefore invisible to the naked eye at so great a distance, they can readily be covered and uncovered since by aid of the telescopes, once adjusted and fixed, they will become easily visible.

SAGR. This experiment strikes me as a clever and reliable invention. But tell us what you conclude from the results.

SALV. In fact I have tried the experiment only at a short distance, less than a mile, from which I have not been able to ascertain with certainty whether the appearance of the opposite

posite light was instantaneous or not; but if not instantaneous
it is extraordinarily rapid—I should call it momentary; and for
the present I should compare it to motion which we see in the
lightning flash between clouds eight or ten miles distant from us.
We see the beginning of this light—I might say its head and
[89]
source—located at a particular place among the clouds; but it
immediately spreads to the surrounding ones, which seems to be
an argument that at least some time is required for propagation;
for if the illumination were instantaneous and not gradual, we
should not be able to distinguish its origin—its center, so to
speak—from its outlying portions. What a sea we are grad-
ually slipping into without knowing it! With vacua and infinities and indivisibles and instantaneous motions, shall we
ever be able, even by means of a thousand discussions, to reach
dry land?

SAGR. Really these matters lie far beyond our grasp. Just
think; when we seek the infinite among numbers we find it in
unity; that which is ever divisible is derived from indivisibles;
the vacuum is found inseparably connected with the plenum;
indeed the views commonly held concerning the nature of these
matters are so reversed that even the circumference of a circle
turns out to be an infinite straight line, a fact which, if my
memory serves me correctly, you, Salviati, were intending to
demonstrate geometrically. Please therefore proceed without
further digression.

SALV. I am at your service; but for the sake of greater clearness let me first demonstrate the following problem:

Given a straight line divided into unequal parts which bear
to each other any ratio whatever, to describe a circle such
that two straight lines drawn from the ends of the given
line to any point on the circumference will bear to each
other the same ratio as the two parts of the given line, thus
making those lines which are drawn from the same terminal
points homologous.

Let AB represent the given straight line divided into any two
unequal parts by the point C; the problem is to describe a circle
such

such that two straight lines drawn from the terminal points, A and B, to any point on the circumference will bear to each other the same ratio as the part AC bears to BC, so that lines drawn from the same terminal points are homologous. About C as center describe a circle having the shorter part CB of the given line, as radius. Through A draw a straight line AD which

[90]

shall be tangent to the circle at D and indefinitely prolonged toward E. Draw the radius CD which will be perpendicular to AE. At B erect a perpendicular to AB; this perpendicular will intersect AE at some point since the angle at A is acute; call this point of intersection E, and from it draw a perpendicular to AE which will intersect AB prolonged in F.

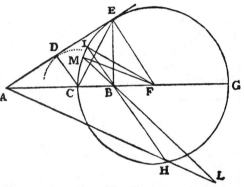

Fig. 8

Now I say the two straight lines FE and FC are equal. For if we join E and C, we shall have two triangles, DEC and BEC, in which the two sides of the one, DE and EC, are equal to the two sides of the other, BE and EC, both DE and EB being tangents to the circle DB while the bases DC and CB are likewise equal; hence the two angles, DEC and BEC, will be equal. Now since the angle BCE differs from a right angle by the angle CEB, and the angle CEF also differs from a right angle by the angle CED, and since these differences are equal, it follows that the angle FCE is equal to CEF; consequently the sides FE and FC are equal. If we describe a circle with F as center and FE as radius it will pass through the point C; let CEG be such a circle. This is the circle sought, for if we draw lines from the terminal points A and B to any point on its circumference they will bear to each other the

the

the same ratio as the two portions AC and BC which meet at the point C. This is manifest in the case of the two lines AE and BE, meeting at the point E, because the angle E of the triangle AEB is bisected by the line CE, and therefore AC: CB = AE: BE. The same may be proved of the two lines AG and BG terminating in the point G. For since the triangles AFE and EFB are similar, we have AF: FE = EF: FB, or AF: FC = CF: FB, and *dividendo* AC: CF = CB: BF, or AC: FG = CB: BF; also *componendo* we have both AB: BG = CB: BF and AG: GB = CF: FB = AE: EB = AC: BC.   Q. E. D.

[91]

Take now any other point in the circumference, say H, where the two lines AH and BH intersect; in like manner we shall have AC: CB = AH: HB. Prolong HB until it meets the circumference at I and join IF; and since we have already found that AB: BG = CB: BF it follows that the rectangle AB.BF is equal to the rectangle CB.BG or IB.BH. Hence AB: BH = IB: BF. But the angles at B are equal and therefore AH: HB = IF: FB = EF: FB = AE: EB.

Besides, I may add, that it is impossible for lines which maintain this same ratio and which are drawn from the terminal points, A and B, to meet at any point either inside or outside the circle, CEG. For suppose this were possible; let AL and BL be two such lines intersecting at the point L outside the circle: prolong LB till it meets the circumference at M and join MF. If AL: BL = AC: BC = MF: FB, then we shall have two triangles ALB and MFB which have the sides about the two angles proportional, the angles at the vertex, B, equal, and the two remaining angles, FMB and LAB, less than right angles (because the right angle at M has for its base the entire diameter CG and not merely a part BF: and the other angle at the point A is acute because the line AL, the homologue of AC, is greater than BL, the homologue of BC). From this it follows that the triangles ABL and MBF are similar and therefore AB: BL = MB: BF, making the rectangle AB.BF = MB.BL; but it has been demonstrated that the rectangle AB.BF is equal to CB.BG; whence it would follow that the rectangle MB.BL is equal to the rectangle

rectangle CB.BG which is impossible; therefore the intersection cannot fall outside the circle. And in like manner we can show that it cannot fall inside; hence all these intersections fall on the circumference.

But now it is time for us to go back and grant the request of Simplicio by showing him that it is not only not impossible to resolve a line into an infinite number of points but that this is quite as easy as to divide it into its finite parts. This I will do under the following condition which I am sure, Simplicio, you will not deny me, namely, that you will not require me to separate the points, one from the other, and show them to you,

[92]

one by one, on this paper; for I should be content that you, without separating the four or six parts of a line from one another, should show me the marked divisions or at most that you should fold them at angles forming a square or a hexagon: for, then, I am certain you would consider the division distinctly and actually accomplished.

SIMP. I certainly should.

SALV. If now the change which takes place when you bend a line at angles so as to form now a square, now an octagon, now a polygon of forty, a hundred or a thousand angles, is sufficient to bring into actuality the four, eight, forty, hundred, and thousand parts which, according to you, existed at first only potentially in the straight line, may I not say, with equal right, that, when I have bent the straight line into a polygon having an infinite number of sides, i. e., into a circle, I have reduced to actuality that infinite number of parts which you claimed, while it was straight, were contained in it only potentially? Nor can one deny that the division into an infinite number of points is just as truly accomplished as the one into four parts when the square is formed or into a thousand parts when the millagon is formed; for in such a division the same conditions are satisfied as in the case of a polygon of a thousand or a hundred thousand sides. Such a polygon laid upon a straight line touches it with one of its sides, i. e., with one of its hundred thousand parts; while the circle which is a polygon of an infinite number of sides touches

touches the same straight line with one of its sides which is a single point different from all its neighbors and therefore separate and distinct in no less degree than is one side of a polygon from the other sides. And just as a polygon, when rolled along a plane, marks out upon this plane, by the successive contacts of its sides, a straight line equal to its perimeter, so the circle rolled upon such a plane also traces by its infinite succession of contacts a straight line equal in length to its own circumference. I am willing, Simplicio, at the outset, to grant to the Peripatetics the truth of their opinion that a continuous quantity [*il continuo*] is divisible only into parts which are still further divisible so that however far the division and subdivision be continued no end will be reached; but I am not so certain that they will concede to me that none of these divisions of theirs can be a final one, as is surely the fact, because there always remains "another"; the final and ultimate division is rather one which resolves a continuous quantity into an infinite number of indivisible quantities, a result which I grant can never be reached by successive division into an ever-increasing number of parts. But if they employ the method which I propose for separating

[93]

and resolving the whole of infinity [*tutta la infinità*], at a single stroke (an artifice which surely ought not to be denied me), I think that they would be contented to admit that a continuous quantity is built up out of absolutely indivisible atoms, especially since this method, perhaps better than any other, enables us to avoid many intricate labyrinths, such as cohesion in solids, already mentioned, and the question of expansion and contraction, without forcing upon us the objectionable admission of empty spaces [in solids] which carries with it the penetrability of bodies. Both of these objections, it appears to me, are avoided if we accept the above-mentioned view of indivisible constituents.

SIMP. I hardly know what the Peripatetics would say since the views advanced by you would strike them as mostly new, and as such we must consider them. It is however not unlikely that they would find answers and solutions for these problems which

I,

I, for want of time and critical ability, am at present unable to solve. Leaving this to one side for the moment, I should like to hear how the introduction of these indivisible quantities helps us to understand contraction and expansion avoiding at the same time the vacuum and the penetrability of bodies.

SAGR. I also shall listen with keen interest to this same matter which is far from clear in my mind; provided I am allowed to hear what, a moment ago, Simplicio suggested we omit, namely, the reasons which Aristotle offers against the existence of the vacuum and the arguments which you must advance in rebuttal.

SALV. I will do both. And first, just as, for the production of expansion, we employ the line described by the small circle during one rotation of the large one—a line greater than the circumference of the small circle—so, in order to explain contraction, we point out that, during each rotation of the smaller circle, the larger one describes a straight line which is shorter than its circumference.

For the better understanding of this we proceed to the consideration of what happens in the case of polygons. Employing

[94]

a figure similar to the earlier one, construct the two hexagons, ABC and HIK, about the common center L, and let them roll along the parallel lines HOM and ABc. Now holding the vertex I fixed, allow the smaller polygon to rotate until the side IK lies upon the parallel, during which motion the point K will describe the arc KM, and the side KI will coincide with IM. Let us see what, in the meantime, the side CB of the larger polygon has been doing. Since the rotation is about the point I, the terminal point B, of the line IB, moving backwards, will describe the arc B*b* underneath the parallel *c*A so that when the side KI coincides with the line MI, the side BC will coincide with *bc*, having advanced only through the distance B*c*, but having retreated through a portion of the line BA which subtends the arc B*b*. If we allow the rotation of the smaller polygon to go on it will traverse and describe along its parallel a line equal to its perimeter; while the larger one will traverse and describe a line less than its perimeter by as many times the length *b*B as there

are

are sides less one; this line is approximately equal to that described by the smaller polygon exceeding it only by the distance bB. Here now we see, without any difficulty, why the larger

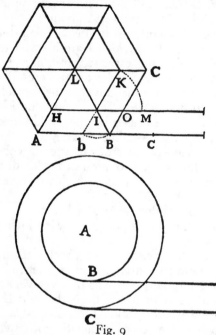

Fig. 9

polygon, when carried by the smaller, does not measure off with its sides a line longer than that traversed by the smaller one; this is because a portion of each side is superposed upon its immediately preceding neighbor.

Let us next consider two circles, having a common center at A, and lying upon their respective parallels, the smaller being tangent to its parallel at the point B; the larger, at the point C. Here when the small circle commences to roll the point B [95] does not remain at rest for a while so as to allow BC to move backward

and carry with it the point C, as happened in the case of the polygons, where the point I remained fixed until the side KI coincided with MI and the line IB carried the terminal point B backward as far as b, so that the side BC fell upon bc, thus superposing upon the line BA, the portion Bb, and advancing by an amount Bc, equal to MI, that is, to one side of the smaller polygon. On account of these superpositions, which are the excesses of the sides of the larger over the smaller polygon, each net advance is equal to one side of the smaller polygon and, during one complete rotation, these amount to a straight line equal in length to the perimeter of the smaller polygon.

But

But now reasoning in the same way concerning the circles, we must observe that whereas the number of sides in any polygon is comprised within a certain limit, the number of sides in a circle is infinite; the former are finite and divisible; the latter infinite and indivisible. In the case of the polygon, the vertices remain at rest during an interval of time which bears to the period of one complete rotation the same ratio which one side bears to the perimeter; likewise, in the case of the circles, the delay of each of the infinite number of vertices is merely instantaneous, because an instant is such a fraction of a finite interval as a point is of a line which contains an infinite number of points. The retrogression of the sides of the larger polygon is not equal to the length of one of its sides but merely to the excess of such a side over one side of the smaller polygon, the net advance being equal to this smaller side; but in the circle, the point or side C, during the instantaneous rest of B, recedes by an amount equal to its excess over the side B, making a net progress equal to B itself. In short the infinite number of indivisible sides of the greater circle with their infinite number of indivisible retrogressions, made during the infinite number of instantaneous delays of the infinite number of vertices of the smaller circle, together with the infinite number of progressions, equal to the infinite number of sides in the smaller circle—all these, I say, add up to a line equal to that described by the smaller circle, a line which contains an infinite number of infinitely small superpositions, thus bringing about a thickening or contraction without any overlapping or interpenetration of finite parts. This result could not be obtained in the case of a line divided

[96]

into finite parts such as is the perimeter of any polygon, which when laid out in a straight line cannot be shortened except by the overlapping and interpenetration of its sides. This contraction of an infinite number of infinitely small parts without the interpenetration or overlapping of finite parts and the previously mentioned [p. 70, Nat. Ed.] expansion of an infinite number of indivisible parts by the interposition of indivisible vacua is, in my opinion, the most that can be said concerning the contraction and

and rarefaction of bodies, unless we give up the impenetrability of matter and introduce empty spaces of finite size. If you find anything here that you consider worth while, pray use it; if not regard it, together with my remarks, as idle talk; but this remember, we are dealing with the infinite and the indivisible.

SAGR. I frankly confess that your idea is subtle and that it impresses me as new and strange; but whether, as a matter of fact, nature actually behaves according to such a law I am unable to determine; however, until I find a more satisfactory explanation I shall hold fast to this one. Perhaps Simplicio can tell us something which I have not yet heard, namely, how to explain the explanation which the philosophers have given of this abstruse matter; for, indeed, all that I have hitherto read concerning contraction is so dense and that concerning expansion so thin that my poor brain can neither penetrate the former nor grasp the latter.

SIMP. I am all at sea and find difficulties in following either path, especially this new one; because according to this theory an ounce of gold might be rarefied and expanded until its size would exceed that of the earth, while the earth, in turn, might be condensed and reduced until it would become smaller than a walnut, something which I do not believe; nor do I believe that you believe it. The arguments and demonstrations which you have advanced are mathematical, abstract, and far removed from concrete matter; and I do not believe that when applied to the physical and natural world these laws will hold.

SALV. I am not able to render the invisible visible, nor do I think that you will ask this. But now that you mention gold, do not our senses tell us that that metal can be immensely expanded? I do not know whether you have observed the method

[97]

employed by those who are skilled in drawing gold wire, of which really only the surface is gold, the inside material being silver. The way they draw it is as follows: they take a cylinder or, if you please, a rod of silver, about half a cubit long and three or four times as wide as one's thumb; this rod they cover with gold-leaf which is so thin that it almost floats in air, putting on

not

not more than eight or ten thicknesses. Once gilded they begin
to pull it, with great force, through the holes of a draw-plate;
again and again it is made to pass through smaller and smaller
holes, until, after very many passages, it is reduced to the
fineness of a lady's hair, or perhaps even finer; yet the surface
remains gilded. Imagine now how the substance of this gold has
been expanded and to what fineness it has been reduced.

SIMP. I do not see that this process would produce, as a
consequence, that marvellous thinning of the substance of the
gold which you suggest: first, because the original gilding con-
sisting of ten layers of gold-leaf has a sensible thickness; secondly,
because in drawing out the silver it grows in length but at the
same time diminishes proportionally in thickness; and, since
one dimension thus compensates the other, the area will not be
so increased as to make it necessary during the process of gilding
to reduce the thinness of the gold beyond that of the original
leaves.

SALV. You are greatly mistaken, Simplicio, because the surface increases directly as the square root of the length, a fact
which I can demonstrate geometrically.

SAGR. Please give us the demonstration not only for my own
sake but also for Simplicio provided you think we can understand it.

SALV. I'll see if I can recall it on the spur of the moment.
At the outset, it is clear that the original thick rod of silver and
the wire drawn out to an enormous length are two cylinders of
the same volume, since they are the same body of silver. So
                              [98]
that, if I determine the ratio between the surfaces of cylinders of
the same volume, the problem will be solved. I say then,

> The areas of cylinders of equal volumes, neglecting the
> bases, bear to each other a ratio which is the square root
> of the ratio of their lengths.

Take two cylinders of equal volume having the altitudes AB
and CD, between which the line E is a mean proportional. Then
I claim that, omitting the bases of each cylinder, the surface of
the cylinder AB is to that of the cylinder CD as the length AB
                                                          is

is to the line E, that is, as the square root of AB is to the square root of CD. Now cut off the cylinder AB at F so that the altitude AF is equal to CD. Then since the bases of cylinders of equal volume bear to one another the inverse ratio of their heights, it follows that the area of the circular base of the cylinder CD will be to the area of the circular base of AB as the altitude BA is to DC: moreover, since circles are to one another as the squares of their diameters, the said squares will be to each other as BA is to CD. But BA is to CD as the square of

Fig. 10

BA is to the square of E: and, therefore, these four squares will form a proportion; and likewise their sides; so the line AB is to E as the diameter of circle C is to the diameter of the circle A. But the diameters are proportional to the circumferences and the circumferences are proportional to the areas of cylinders of equal height; hence the line AB is to E as the surface of the cylinder CD is to the surface of the cylinder AF. Now since the height AF is to AB as the surface of AF is to the surface of AB; and since the height AB is to the line E as the surface CD is to AF, it follows, *ex æquali in proportione perturbata*,* that the height AF is to E as the surface CD is to the surface AB, and *convertendo*, the surface of the cylinder AB is to the surface of the cylinder CD as the line E is to AF, i. e., to CD, or as AB is to E which is the square root of the ratio of AB to CD. Q. E. D.

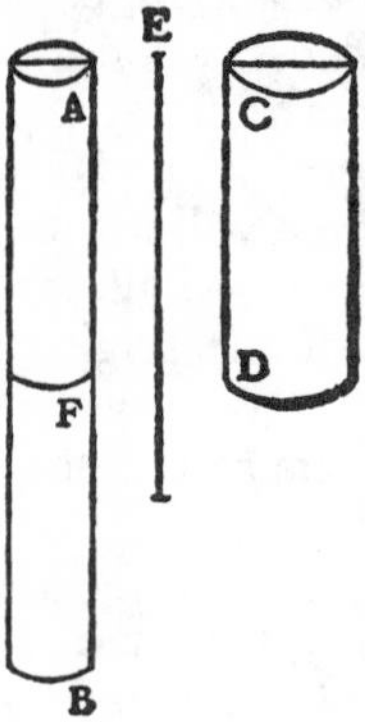

If now we apply these results to the case in hand, and assume that the silver cylinder at the time of gilding had a length of only half a cubit and a thickness three or four times that of

[99]

one's thumb, we shall find that, when the wire has been reduced to the fineness of a hair and has been drawn out to a length of twenty thousand cubits (and perhaps more), the area of its surface will have been increased not less than two hundred times. Consequently the ten leaves of gold which were laid on

* See *Euclid*, Book V, Def. 20., Todhunter's Ed., p. 137 (London, 1877.)
[*Trans.*]

have been extended over a surface two hundred times greater, assuring us that the thickness of the gold which now covers the surface of so many cubits of wire cannot be greater than one twentieth that of an ordinary leaf of beaten gold. Consider now what degree of fineness it must have and whether one could conceive it to happen in any other way than by enormous expansion of parts; consider also whether this experiment does not suggest that physical bodies [*materie fisiche*] are composed of infinitely small indivisible particles, a view which is supported by other more striking and conclusive examples.

SAGR. This demonstration is so beautiful that, even if it does not have the cogency originally intended,—although to my mind, it is very forceful—the short time devoted to it has nevertheless been most happily spent.

SALV. Since you are so fond of these geometrical demonstrations, which carry with them distinct gain, I will give you a companion theorem which answers an extremely interesting query. We have seen above what relations hold between equal cylinders of different height or length; let us now see what holds when the cylinders are equal in area but unequal in height, understanding area to include the curved surface, but not the upper and lower bases. The theorem is:

> The volumes of right cylinders having equal curved surfaces are inversely proportional to their altitudes.

Let the surfaces of the two cylinders, AE and CF, be equal but let the height of the latter, CD, be greater than that of the former, AB: then I say that the volume of the cylinder AE is to that of the cylinder CF as the height CD is to AB. Now since the surface of CF is equal to the surface of AE, it follows that the volume of CF is less than that of AE; for, if they were equal, the surface of CF would, by the preceding proposition, exceed that of AE, and the excess would be so much the greater if the volume of the cylinder CF were greater than that

[100]

of AE. Let us now take a cylinder ID having a volume equal to that of AE; then, according to the preceding theorem, the surface of the cylinder ID is to the surface of AE as the altitude

IF

IF is to the mean proportional between IF and AB. But since one datum of the problem is that the surface of AE is equal to that of CF, and since the surface ID is to the surface CF as the altitude IF is to the altitude CD, it follows that CD is a mean proportional between IF and AB. Not only so, but since the volume of the cylinder ID is equal to that of AE, each will bear the same ratio to the volume of the cylinder CF; but the volume ID is to the volume CF as the altitude IF is to the altitude CD; hence the volume of AE is to the volume of CF as the length IF is to the length CD, that is, as the length CD is to the length AB.   Q. E. D.

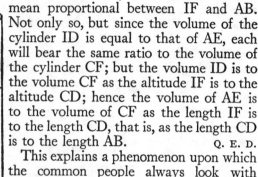

Fig. 11

This explains a phenomenon upon which the common people always look with wonder, namely, if we have a piece of stuff which has one side longer than the other, we can make from it a cornsack, using the customary wooden base, which will hold more when the short side of the cloth is used for the height of the sack and the long side is wrapped around the wooden base, than with the alternative arrangement. So that, for instance, from a piece of cloth which is six cubits on one side and twelve on the other, a sack can be made which will hold more when the side of twelve cubits is wrapped around the wooden base, leaving the sack six cubits high than when the six cubit side is put around the base making the sack twelve cubits high. From what has been proven above we learn not only the general fact that one sack holds more than the other, but we also get specific and particular information as to how much more, namely, just in proportion as the altitude of the sack diminishes the contents increase and *vice versa*. Thus if we use the figures given which make the cloth twice as long as wide and if we use the long side for the seam, the volume of the sack will be just one-half as great as with the opposite arrangement. Likewise if

[101]

if we have a piece of matting which measures 7 x 25 cubits and make from it a basket, the contents of the basket will, when the seam is lengthwise, be seven as compared with twenty-five when the seam runs endwise.

Sagr. It is with great pleasure that we continue thus to acquire new and useful information. But as regards the subject just discussed, I really believe that, among those who are not already familiar with geometry, you would scarcely find four persons in a hundred who would not, at first sight, make the mistake of believing that bodies having equal surfaces would be equal in other respects. Speaking of areas, the same error is made when one attempts, as often happens, to determine the sizes of various cities by measuring their boundary lines, forgetting that the circuit of one may be equal to the circuit of another while the area of the one is much greater than that of the other. And this is true not only in the case of irregular, but also of regular surfaces, where the polygon having the greater number of sides always contains a larger area than the one with the less number of sides, so that finally the circle which is a polygon of an infinite number of sides contains the largest area of all polygons of equal perimeter. I remember with particular pleasure having seen this demonstration when I was studying the sphere of Sacrobosco * with the aid of a learned commentary.

Salv. Very true! I too came across the same passage which suggested to me a method of showing how, by a single short demonstration, one can prove that the circle has the largest content of all regular isoperimetric figures; and that, of other

[102]

figures, the one which has the larger number of sides contains a greater area than that which has the smaller number.

Sagr. Being exceedingly fond of choice and uncommon propositions, I beseech you to let us have your demonstration.

Salv. I can do this in a few words by proving the following theorem:

The area of a circle is a mean proportional between any

* See interesting biographical note on Sacrobosco [John Holywood] in *Ency. Brit.*, 11th Ed. [*Trans.*]

two regular and similar polygons of which one circumscribes it and the other is isoperimetric with it. In addition,
the area of the circle is less than that of any circumscribed
polygon and greater than that of any isoperimetric polygon.
And further, of these circumscribed polygons, the one which
has the greater number of sides is smaller than the one which
has a less number; but, on the other hand, that isoperimetric polygon which has the greater number of sides is
the larger.

   Let A and B be two similar polygons of which A circumscribes
the given circle and B is isoperimetric with it. The area of the
circle will then be a mean proportional between the areas of the
polygons. For if we indicate the radius of the circle by AC and
if we remember that the area of the circle is equal to that of a
right-angled triangle in which one of the sides about the right
angle is equal to the radius, AC, and the other to the circumference; and if likewise we remember that the area of the polygon A is equal to the area of a right-angled triangle one of
                              [103]
whose sides about the right angle has the same length as AC and
the other is equal to the perimeter of the polygon itself; it is then

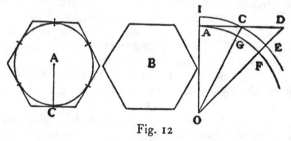

Fig. 12

manifest that the circumscribed polygon bears to the circle the
same ratio which its perimeter bears to the circumference of the
circle, or to the perimeter of the polygon B which is, by hypothesis, equal to the circumference of the circle. But since the
polygons A and B are similar their areas are to each other as the
squares of their perimeters; hence the area of the circle A is a
                                                          mean

mean proportional between the areas of the two polygons A and B. And since the area of the polygon A is greater than that of the circle A, it is clear that the area of the circle A is greater than that of the isoperimetric polygon B, and is therefore the greatest of all regular polygons having the same perimeter as the circle.

We now demonstrate the remaining portion of the theorem, which is to prove that, in the case of polygons circumscribing a given circle, the one having the smaller number of sides has a larger area than one having a greater number of sides; but that on the other hand, in the case of isoperimetric polygons, the one having the more sides has a larger area than the one with less sides. To the circle which has O for center and OA for radius draw the tangent AD; and on this tangent lay off, say, AD which shall represent one-half of the side of a circumscribed pentagon and AC which shall represent one-half of the side of a heptagon; draw the straight lines OGC and OFD; then with O as a center and OC as radius draw the arc ECI. Now since the triangle DOC is greater than the sector EOC and since the sector COI is greater than the triangle COA, it follows that the triangle DOC bears to the triangle COA a greater ratio than the sector EOC bears to the sector COI, that is, than the sector FOG bears to the sector GOA. Hence, *componendo et permutando*, the triangle DOA bears to the sector FOA a greater ratio than that which the triangle COA bears to the sector GOA, and also 10 such triangles DOA bear to 10 such sectors FOA a greater ratio than 14 such triangles COA bear to 14 such sectors GOA, that is to say, the circumscribed pentagon bears to the circle a greater ratio than does the heptagon. Hence the pentagon exceeds the heptagon in area.

But now let us assume that both the heptagon and the pentagon have the same perimeter as that of a given circle. Then I say the heptagon will contain a larger area than the pentagon. For since the area of the circle is a mean proportional between areas of the circumscribed and of the isoperimetric pentagons,

[104]

and since likewise it is a mean proportional between the circumscribed

cumscribed and isoperimetric heptagons, and since also we have proved that the circumscribed pentagon is larger than the circumscribed heptagon, it follows that this circumscribed pentagon bears to the circle a larger ratio than does the heptagon, that is, the circle will bear to its isoperimetric pentagon a greater ratio than to its isoperimetric heptagon. Hence the pentagon is smaller than its isoperimetric heptagon. Q. E. D.

SAGR. A very clever and elegant demonstration! But how did we come to plunge into geometry while discussing the objections urged by Simplicio, objections of great moment, especially that one referring to density which strikes me as particularly difficult?

SALV. If contraction and expansion [*condensazione e rarefazzione*] consist in contrary motions, one ought to find for each great expansion a correspondingly large contraction. But our surprise is increased when, every day, we see enormous expansions taking place almost instantaneously. Think what a tremendous expansion occurs when a small quantity of gunpowder flares up into a vast volume of fire! Think too of the almost limitless expansion of the light which it produces! Imagine the contraction which would take place if this fire and this light were to reunite, which, indeed, is not impossible since only a little while ago they were located together in this small space. You will find, upon observation, a thousand such expansions for they are more obvious than contractions since dense matter is more palpable and accessible to our senses. We can take wood and see it go up in fire and light, but we do not see

[105]

them recombine to form wood; we see fruits and flowers and a thousand other solid bodies dissolve largely into odors, but we do not observe these fragrant atoms coming together to form fragrant solids. But where the senses fail us reason must step in; for it will enable us to understand the motion involved in the condensation of extremely rarefied and tenuous substances just as clearly as that involved in the expansion and dissolution of solids. Moreover we are trying to find out how it is possible to produce expansion and contraction in bodies which are capable of such changes without introducing vacua and without giving

up

up the impenetrability of matter; but this does not exclude the possibility of there being materials which possess no such properties and do not, therefore, carry with them consequences which you call inconvenient and impossible. And finally, Simplicio, I have, for the sake of you philosophers, taken pains to find an explanation of how expansion and contraction can take place without our admitting the penetrability of matter and introducing vacua, properties which you deny and dislike; if you were to admit them, I should not oppose you so vigorously. Now either admit these difficulties or accept my views or suggest something better.

SAGR. I quite agree with the peripatetic philosophers in denying the penetrability of matter. As to the vacua I should like to hear a thorough discussion of Aristotle's demonstration in which he opposes them, and what you, Salviati, have to say in reply. I beg of you, Simplicio, that you give us the precise proof of the Philosopher and that you, Salviati, give us the reply.

SIMP. So far as I remember, Aristotle inveighs against the ancient view that a vacuum is a necessary prerequisite for motion and that the latter could not occur without the former. In opposition to this view Aristotle shows that it is precisely the phenomenon of motion, as we shall see, which renders untenable the idea of a vacuum. His method is to divide the argument into two parts. He first supposes bodies of different weights to move in the same medium; then supposes, one and the same body to move in different media. In the first case, he

[106]

supposes bodies of different weight to move in one and the same medium with different speeds which stand to one another in the same ratio as the weights; so that, for example, a body which is ten times as heavy as another will move ten times as rapidly as the other. In the second case he assumes that the speeds of one and the same body moving in different media are in inverse ratio to the densities of these media; thus, for instance, if the density of water were ten times that of air, the speed in air would be ten times greater than in water. From this second supposition,

tion, he shows that, since the tenuity of a vacuum differs infinitely from that of any medium filled with matter however rare, any body which moves in a plenum through a certain space in a certain time ought to move through a vacuum instantaneously; but instantaneous motion is an impossibility; it is therefore impossible that a vacuum should be produced by motion.

SALV. The argument is, as you see, *ad hominem*, that is, it is directed against those who thought the vacuum a prerequisite for motion. Now if I admit the argument to be conclusive and concede also that motion cannot take place in a vacuum, the assumption of a vacuum considered absolutely and not with reference to motion, is not thereby invalidated. But to tell you what the ancients might possibly have replied and in order to better understand just how conclusive Aristotle's demonstration is, we may, in my opinion, deny both of his assumptions. And as to the first, I greatly doubt that Aristotle ever tested by experiment whether it be true that two stones, one weighing ten times as much as the other, if allowed to fall, at the same instant, from a height of, say, 100 cubits, would so differ in speed that when the heavier had reached the ground, the other would not have fallen more than 10 cubits.

SIMP. His language would seem to indicate that he had tried the experiment, because he says: *We see the heavier;* now the word *see* shows that he had made the experiment.

SAGR. But I, Simplicio, who have made the test can assure
[107]
you that a cannon ball weighing one or two hundred pounds, or even more, will not reach the ground by as much as a span ahead of a musket ball weighing only half a pound, provided both are dropped from a height of 200 cubits.

SALV. But, even without further experiment, it is possible to prove clearly, by means of a short and conclusive argument, that a heavier body does not move more rapidly than a lighter one provided both bodies are of the same material and in short such as those mentioned by Aristotle. But tell me, Simplicio, whether you admit that each falling body acquires a definite speed

speed fixed by nature, a velocity which cannot be increased or diminished except by the use of force [*violenza*] or resistance.

SIMP. There can be no doubt but that one and the same body moving in a single medium has a fixed velocity which is determined by nature and which cannot be increased except by the addition of momentum [*impeto*] or diminished except by some resistance which retards it.

SALV. If then we take two bodies whose natural speeds are different, it is clear that on uniting the two, the more rapid one will be partly retarded by the slower, and the slower will be somewhat hastened by the swifter. Do you not agree with me in this opinion?

SIMP. You are unquestionably right.

SALV. But if this is true, and if a large stone moves with a speed of, say, eight while a smaller moves with a speed of four, then when they are united, the system will move with a speed less than eight; but the two stones when tied together make a stone larger than that which before moved with a speed of eight. Hence the heavier body moves with less speed than the lighter; an effect which is contrary to your supposition. Thus you see

[108]

how, from your assumption that the heavier body moves more rapidly than the lighter one, I infer that the heavier body moves more slowly.

SIMP. I am all at sea because it appears to me that the smaller stone when added to the larger increases its weight and by adding weight I do not see how it can fail to increase its speed or, at least, not to diminish it.

SALV. Here again you are in error, Simplicio, because it is not true that the smaller stone adds weight to the larger.

SIMP. This is, indeed, quite beyond my comprehension.

SALV. It will not be beyond you when I have once shown you the mistake under which you are laboring. Note that it is necessary to distinguish between heavy bodies in motion and the same bodies at rest. A large stone placed in a balance not only acquires additional weight by having another stone placed upon it, but even by the addition of a handful of hemp its weight is augmented

augmented six to ten ounces according to the quantity of hemp. But if you tie the hemp to the stone and allow them to fall freely from some height, do you believe that the hemp will press down upon the stone and thus accelerate its motion or do you think the motion will be retarded by a partial upward pressure? One always feels the pressure upon his shoulders when he prevents the motion of a load resting upon him; but if one descends just as rapidly as the load would fall how can it gravitate or press upon him? Do you not see that this would be the same as trying to strike a man with a lance when he is running away from you with a speed which is equal to, or even greater, than that with which you are following him? You must therefore conclude that, during free and natural fall, the small stone does not press upon the larger and consequently does not increase its weight as it does when at rest.

SIMP. But what if we should place the larger stone upon the smaller?

[109]

SALV. Its weight would be increased if the larger stone moved more rapidly; but we have already concluded that when the small stone moves more slowly it retards to some extent the speed of the larger, so that the combination of the two, which is a heavier body than the larger of the two stones, would move less rapidly, a conclusion which is contrary to your hypothesis. We infer therefore that large and small bodies move with the same speed provided they are of the same specific gravity.

SIMP. Your discussion is really admirable; yet I do not find it easy to believe that a bird-shot falls as swiftly as a cannon ball.

SALV. Why not say a grain of sand as rapidly as a grindstone? But, Simplicio, I trust you will not follow the example of many others who divert the discussion from its main intent and fasten upon some statement of mine which lacks a hair's-breadth of the truth and, under this hair, hide the fault of another which is as big as a ship's cable. Aristotle says that "an iron ball of one hundred pounds falling from a height of one hundred cubits reaches the ground before a one-pound ball has fallen a single cubit." I say that they arrive at the same time. You find, on
making

making the experiment, that the larger outstrips the smaller by two finger-breadths, that is, when the larger has reached the ground, the other is short of it by two finger-breadths; now you would not hide behind these two fingers the ninety-nine cubits of Aristotle, nor would you mention my small error and at the same time pass over in silence his very large one. Aristotle declares that bodies of different weights, in the same medium, travel (in so far as their motion depends upon gravity) with speeds which are proportional to their weights; this he illustrates by use of bodies in which it is possible to perceive the pure and un-adulterated effect of gravity, eliminating other considerations, for example, figure as being of small importance [*minimi momenti*], influences which are greatly dependent upon the medium which modifies the single effect of gravity alone. Thus we observe that gold, the densest of all substances, when beaten out into a very thin leaf, goes floating through the air; the same thing happens with stone when ground into a very fine powder. But if you wish to maintain the general proposition you will have to show that the same ratio of speeds is preserved in the

[110]

case of all heavy bodies, and that a stone of twenty pounds moves ten times as rapidly as one of two; but I claim that this is false and that, if they fall from a height of fifty or a hundred cubits, they will reach the earth at the same moment.

SIMP. Perhaps the result would be different if the fall took place not from a few cubits but from some thousands of cubits.

SALV. If this were what Aristotle meant you would burden him with another error which would amount to a falsehood; because, since there is no such sheer height available on earth, it is clear that Aristotle could not have made the experiment; yet he wishes to give us the impression of his having performed it when he speaks of such an effect as one which we see.

SIMP. In fact, Aristotle does not employ this principle, but uses the other one which is not, I believe, subject to these same difficulties.

SALV. But the one is as false as the other; and I am surprised that you yourself do not see the fallacy and that you do not

perceive

perceive that if it were true that, in media of different densities
and different resistances, such as water and air, one and the
same body moved in air more rapidly than in water, in proportion as the density of water is greater than that of air, then it
would follow that any body which falls through air ought also
to fall through water.   But this conclusion is false inasmuch as
many bodies which descend in air not only do not descend in
water, but actually rise.

SIMP. I do not understand the necessity of your inference;
and in addition I will say that Aristotle discusses only those
bodies which fall in both media, not those which fall in air but
rise in water.

SALV. The arguments which you advance for the Philosopher are such as he himself would have certainly avoided so as
not to aggravate his first mistake.   But tell me now whether the
density [*corpulenza*] of the water, or whatever it may be that

[111]

retards the motion, bears a definite ratio to the density of air
which is less retardative; and if so fix a value for it at your
pleasure.

SIMP. Such a ratio does exist; let us assume it to be ten; then,
for a body which falls in both these media, the speed in water will
be ten times slower than in air.

SALV. I shall now take one of those bodies which fall in air
but not in water, say a wooden ball, and I shall ask you to assign
to it any speed you please for its descent through air.

SIMP. Let us suppose it moves with a speed of twenty.

SALV. Very well.   Then it is clear that this speed bears to
some smaller speed the same ratio as the density of water bears
to that of air; and the value of this smaller speed is two.   So
that really if we follow exactly the assumption of Aristotle we
ought to infer that the wooden ball which falls in air, a substance ten times less-resisting than water, with a speed of twenty
would fall in water with a speed of two, instead of coming to the
surface from the bottom as it does; unless perhaps you wish to
reply, which I do not believe you will, that the rising of the wood
through the water is the same as its falling with a speed of two.

But

But since the wooden ball does not go to the bottom, I think you will agree with me that we can find a ball of another material, not wood, which does fall in water with a speed of two.

SIMP. Undoubtedly we can; but it must be of a substance considerably heavier than wood.

SALV. That is it exactly. But if this second ball falls in water with a speed of two, what will be its speed of descent in air? If you hold to the rule of Aristotle you must reply that it will move at the rate of twenty; but twenty is the speed which you yourself have already assigned to the wooden ball; hence this and the other heavier ball will each move through air with the same speed. But now how does the Philosopher harmonize this result with his other, namely, that bodies of different weight move through the same medium with different speeds—speeds which are proportional to their weights? But without going into the matter more deeply, how have these common and

[112]

obvious properties escaped your notice? Have you not observed that two bodies which fall in water, one with a speed a hundred times as great as that of the other, will fall in air with speeds so nearly equal that one will not surpass the other by as much as one hundredth part? Thus, for example, an egg made of marble will descend in water one hundred times more rapidly than a hen's egg, while in air falling from a height of twenty cubits the one will fall short of the other by less than four finger-breadths. In short, a heavy body which sinks through ten cubits of water in three hours will traverse ten cubits of air in one or two pulse-beats; and if the heavy body be a ball of lead it will easily traverse the ten cubits of water in less than double the time required for ten cubits of air. And here, I am sure, Simplicio, you find no ground for difference or objection. We conclude, therefore, that the argument does not bear against the existence of a vacuum; but if it did, it would only do away with vacua of considerable size which neither I nor, in my opinion, the ancients ever believed to exist in nature, although they might possibly be produced by force [*violenza*] as may be gathered from various experiments whose description would here occupy too much time.

Sagr.

SAGR. Seeing that Simplicio is silent, I will take the opportunity of saying something. Since you have clearly demonstrated that bodies of different weights do not move in one and the same medium with velocities proportional to their weights, but that they all move with the same speed, understanding of course that they are of the same substance or at least of the same specific gravity; certainly not of different specific gravities, for I hardly think you would have us believe a ball of cork moves

[113]

with the same speed as one of lead; and again since you have clearly demonstrated that one and the same body moving through differently resisting media does not acquire speeds which are inversely proportional to the resistances, I am curious to learn what are the ratios actually observed in these cases.

SALV. These are interesting questions and I have thought much concerning them. I will give you the method of approach and the result which I finally reached. Having once established the falsity of the proposition that one and the same body moving through differently resisting media acquires speeds which are inversely proportional to the resistances of these media, and having also disproved the statement that in the same medium bodies of different weight acquire velocities proportional to their weights (understanding that this applies also to bodies which differ merely in specific gravity), I then began to combine these two facts and to consider what would happen if bodies of different weight were placed in media of different resistances; and I found that the differences in speed were greater in those media which were more resistant, that is, less yielding. This difference was such that two bodies which differed scarcely at all in their speed through air would, in water, fall the one with a speed ten times as great as that of the other. Further, there are bodies which will fall rapidly in air, whereas if placed in water not only will not sink but will remain at rest or will even rise to the top: for it is possible to find some kinds of wood, such as knots and roots, which remain at rest in water but fall rapidly in air.

SAGR. I have often tried with the utmost patience to add grains of sand to a ball of wax until it should acquire the same

specific

specific gravity as water and would therefore remain at rest in this medium. But with all my care I was never able to accomplish this. Indeed, I do not know whether there is any solid substance whose specific gravity is, by nature, so nearly equal to that of water that if placed anywhere in water it will remain at rest.

SALV. In this, as in a thousand other operations, men are surpassed by animals. In this problem of yours one may learn much from the fish which are very skillful in maintaining their equilibrium not only in one kind of water, but also in waters which are notably different either by their own nature or by

[114]

some accidental muddiness or through salinity, each of which produces a marked change. So perfectly indeed can fish keep their equilibrium that they are able to remain motionless in any position. This they accomplish, I believe, by means of an apparatus especially provided by nature, namely, a bladder located in the body and communicating with the mouth by means of a narrow tube through which they are able, at will, to expel a portion of the air contained in the bladder: by rising to the surface they can take in more air; thus they make themselves heavier or lighter than water at will and maintain equilibrium.

SAGR. By means of another device I was able to deceive some friends to whom I had boasted that I could make up a ball of wax that would be in equilibrium in water. In the bottom of a vessel I placed some salt water and upon this some fresh water; then I showed them that the ball stopped in the middle of the water, and that, when pushed to the bottom or lifted to the top, would not remain in either of these places but would return to the middle.

SALV. This experiment is not without usefulness. For when physicians are testing the various qualities of waters, especially their specific gravities, they employ a ball of this kind so adjusted that, in certain water, it will neither rise nor fall. Then in testing another water, differing ever so slightly in specific gravity [peso], the ball will sink if this water be lighter and rise if it be heavier. And so exact is this experiment that the addition

tion

tion of two grains of salt to six pounds of water is sufficient to make the ball rise to the surface from the bottom to which it had fallen. To illustrate the precision of this experiment and also to clearly demonstrate the non-resistance of water to division, I wish to add that this notable difference in specific gravity can be produced not only by solution of some heavier substance, but also by merely heating or cooling; and so sensitive is water to this process that by simply adding four drops of another water which is slightly warmer or cooler than the six pounds one can cause the ball to sink or rise; it will sink when the warm water is poured in and will rise upon the addition of cold water. Now you

[115]

can see how mistaken are those philosophers who ascribe to water viscosity or some other coherence of parts which offers resistance to separation of parts and to penetration.

SAGR. With regard to this question I have found many convincing arguments in a treatise by our Academician; but there is one great difficulty of which I have not been able to rid myself, namely, if there be no tenacity or coherence between the particles of water how is it possible for those large drops of water to stand out in relief upon cabbage leaves without scattering or spreading out?

SALV. Although those who are in possession of the truth are able to solve all objections raised, I would not arrogate to myself such power; nevertheless my inability should not be allowed to becloud the truth. To begin with let me confess that I do not understand how these large globules of water stand out and hold themselves up, although I know for a certainty, that it is not owing to any internal tenacity acting between the particles of water; whence it must follow that the cause of this effect is external. Beside the experiments already shown to prove that the cause is not internal, I can offer another which is very convincing. If the particles of water which sustain themselves in a heap, while surrounded by air, did so in virtue of an internal cause then they would sustain themselves much more easily when surrounded by a medium in which they exhibit less tendency to fall than they do in air; such a medium would be any fluid
heavier

heavier than air, as, for instance, wine: and therefore if some wine be poured about such a drop of water, the wine might rise until the drop was entirely covered, without the particles of water, held together by this internal coherence, ever parting company. But this is not the fact; for as soon as the wine touches the water, the latter without waiting to be covered scatters and spreads out underneath the wine if it be red. The cause of this effect is therefore external and is possibly to be found in the surrounding air. Indeed there appears to be a considerable antagonism between air and water as I have observed in the following experiment. Having taken a glass globe which had a mouth of about the same diameter as a straw, I filled it with water and turned it mouth downwards; neverthe-

[116]

less, the water, although quite heavy and prone to descend, and the air, which is very light and disposed to rise through the water, refused, the one to descend and the other to ascend through the opening, but both remained stubborn and defiant. On the other hand, as soon as I apply to this opening a glass of red wine, which is almost inappreciably lighter than water, red streaks are immediately observed to ascend slowly through the water while the water with equal slowness descends through the wine without mixing, until finally the globe is completely filled with wine and the water has all gone down into the vessel below. What then can we say except that there exists, between water and air, a certain incompatibility which I do not understand, but perhaps. . . .

SIMP. I feel almost like laughing at the great antipathy which Salviati exhibits against the use of the word antipathy; and yet it is excellently adapted to explain the difficulty.

SALV. Alright, if it please Simplicio, let this word antipathy be the solution of our difficulty. Returning from this digression, let us again take up our problem. We have already seen that the difference of speed between bodies of different specific gravities is most marked in those media which are the most resistant: thus, in a medium of quicksilver, gold not merely sinks to the bottom more rapidly than lead but it is the only substance

substance that will descend at all; all other metals and stones rise to the surface and float.   On the other hand the variation of speed in air between balls of gold, lead, copper, porphyry, and other heavy materials is so slight that in a fall of 100 cubits a ball of gold would surely not outstrip one of copper by as much as four fingers.   Having observed this I came to the conclusion that in a medium totally devoid of resistance all bodies would fall with the same speed.

SIMP. This is a remarkable statement, Salviati.   But I shall never believe that even in a vacuum, if motion in such a place were possible, a lock of wool and a bit of lead can fall with the same velocity.

SALV. A little more slowly, Simplicio.   Your difficulty is not so recondite nor am I so imprudent as to warrant you in believing that I have not already considered this matter and found the proper solution.   Hence for my justification and

[117]

for your enlightenment hear what I have to say.   Our problem is to find out what happens to bodies of different weight moving in a medium devoid of resistance, so that the only difference in speed is that which arises from inequality of weight.   Since no medium except one entirely free from air and other bodies, be it ever so tenuous and yielding, can furnish our senses with the evidence we are looking for, and since such a medium is not available, we shall observe what happens in the rarest and least resistant media as compared with what happens in denser and more resistant media.   Because if we find as a fact that the variation of speed among bodies of different specific gravities is less and less according as the medium becomes more and more yielding, and if finally in a medium of extreme tenuity, though not a perfect vacuum, we find that, in spite of great diversity of specific gravity [*peso*], the difference in speed is very small and almost inappreciable, then we are justified in believing it highly probable that in a vacuum all bodies would fall with the same speed.   Let us, in view of this, consider what takes place in air, where for the sake of a definite figure and light material imagine an inflated bladder.   The air in this bladder when surrounded by

air

air will weigh little or nothing, since it can be only slightly compressed; its weight then is small being merely that of the skin which does not amount to the thousandth part of a mass of lead having the same size as the inflated bladder. Now, Simplicio, if we allow these two bodies to fall from a height of four or six cubits, by what distance do you imagine the lead will anticipate the bladder? You may be sure that the lead will not travel three times, or even twice, as swiftly as the bladder, although you would have made it move a thousand times as rapidly.

SIMP. It may be as you say during the first four or six cubits of the fall; but after the motion has continued a long while, I believe that the lead will have left the bladder behind not only six out of twelve parts of the distance but even eight or ten.

SALV. I quite agree with you and doubt not that, in very long distances, the lead might cover one hundred miles while the

[118]

bladder was traversing one; but, my dear Simplicio, this phenomenon which you adduce against my proposition is precisely the one which confirms it. Let me once more explain that the variation of speed observed in bodies of different specific gravities is not caused by the difference of specific gravity but depends upon external circumstances and, in particular, upon the resistance of the medium, so that if this is removed all bodies would fall with the same velocity; and this result I deduce mainly from the fact which you have just admitted and which is very true, namely, that, in the case of bodies which differ widely in weight, their velocities differ more and more as the spaces traversed increase, something which would not occur if the effect depended upon differences of specific gravity. For since these specific gravities remain constant, the ratio between the distances traversed ought to remain constant whereas the fact is that this ratio keeps on increasing as the motion continues. Thus a very heavy body in a fall of one cubit will not anticipate a very light one by so much as the tenth part of this space; but in a fall of twelve cubits the heavy body would outstrip

strip the other by one-third, and in a fall of one hundred cubits by 90/100, etc.

Simp. Very well: but, following your own line of argument, if differences of weight in bodies of different specific gravities cannot produce a change in the ratio of their speeds, on the ground that their specific gravities do not change, how is it possible for the medium, which also we suppose to remain constant, to bring about any change in the ratio of these velocities?

Salv. This objection with which you oppose my statement is clever; and I must meet it. I begin by saying that a heavy body has an inherent tendency to move with a constantly and uniformly accelerated motion toward the common center of gravity, that is, toward the center of our earth, so that during equal intervals of time it receives equal increments of momentum and velocity. This, you must understand, holds whenever all external and accidental hindrances have been removed; but of these there is one which we can never remove, namely, the medium which must be penetrated and thrust aside by the falling body. This quiet, yielding, fluid medium opposes motion

[119]

through it with a resistance which is proportional to the rapidity with which the medium must give way to the passage of the body; which body, as I have said, is by nature continuously accelerated so that it meets with more and more resistance in the medium and hence a diminution in its rate of gain of speed until finally the speed reaches such a point and the resistance of the medium becomes so great that, balancing each other, they prevent any further acceleration and reduce the motion of the body to one which is uniform and which will thereafter maintain a constant value. There is, therefore, an increase in the resistance of the medium, not on account of any change in its essential properties, but on account of the change in rapidity with which it must yield and give way laterally to the passage of the falling body which is being constantly accelerated.

Now seeing how great is the resistance which the air offers to the slight momentum [*momento*] of the bladder and how small that which it offers to the large weight [*peso*] of the lead, I am am convinced that, if the medium were entirely removed, the advantage received by the bladder would be so great and that coming to the lead so small that their speeds would be equalized. Assuming this principle, that all falling bodies acquire equal speeds in a medium which, on account of a vacuum or something else, offers no resistance to the speed of the motion, we shall be able accordingly to determine the ratios of the speeds of both similar and dissimilar bodies moving either through one and the same medium or through different space-filling, and therefore resistant, media. This result we may obtain by observing how much the weight of the medium detracts from the weight of the moving body, which weight is the means employed by the falling body to open a path for itself and to push aside the parts of the medium, something which does not happen in a vacuum where, therefore, no difference [of speed] is to be expected from a difference of specific gravity. And since it is known that the effect of the medium is to diminish the weight of the body by the weight of the medium displaced, we may accomplish our purpose by diminishing in just this proportion the speeds of the falling bodies, which in a non-resisting medium we have assumed to be equal.

Thus, for example, imagine lead to be ten thousand times as heavy as air while ebony is only one thousand times as heavy.

[120]

Here we have two substances whose speeds of fall in a medium devoid of resistance are equal: but, when air is the medium, it will subtract from the speed of the lead one part in ten thousand, and from the speed of the ebony one part in one thousand, i. e. ten parts in ten thousand. While therefore lead and ebony would fall from any given height in the same interval of time, provided the retarding effect of the air were removed, the lead will, in air, lose in speed one part in ten thousand; and the ebony, ten parts in.ten thousand. In other words, if the elevation from which the bodies start be divided into ten thousand parts, the lead will reach the ground leaving the ebony behind by as much as ten, or at least nine, of these parts. Is it not clear then that a leaden ball allowed to fall from a tower two hundred cubits
high

high will outstrip an ebony ball by less than four inches? Now ebony weighs a thousand times as much as air but this inflated bladder only four times as much; therefore air diminishes the inherent and natural speed of ebony by one part in a thousand; while that of the bladder which, if free from hindrance, would be the same, experiences a diminution in air amounting to one part in four.   So that when the ebony ball, falling from the tower, has reached the earth, the bladder will have traversed only three-quarters of this distance.   Lead is twelve times as heavy as water; but ivory is only twice as heavy.   The speeds of these two substances which, when entirely unhindered, are equal will be diminished in water, that of lead by one part in twelve, that of ivory by half.   Accordingly when the lead has fallen through eleven cubits of water the ivory will have fallen through only six.   Employing this principle we shall, I believe, find a much closer agreement of experiment with our computation than with that of Aristotle.

In a similar manner we may find the ratio of the speeds of one and the same body in different fluid media, not by comparing the different resistances of the media, but by considering the excess of the specific gravity of the body above those of the media. Thus, for example, tin is one thousand times heavier than air and ten times heavier than water; hence, if we divide its unhindered speed into 1000 parts, air will rob it of one of these parts so that it will fall with a speed of 999, while in water its speed will be 900, seeing that water diminishes its weight by one part in ten while air by only one part in a thousand.

Again take a solid a little heavier than water, such as oak, a ball of which will weigh let us say 1000 drachms; suppose an

[121]

equal volume of water to weigh 950, and an equal volume of air, 2; then it is clear that if the unhindered speed of the ball is 1000, its speed in air will be 998, but in water only 50, seeing that the water removes 950 of the 1000 parts which the body weighs, leaving only 50.

Such a solid would therefore move almost twenty times as fast in air as in water, since its specific gravity exceeds that of

water

water by one part in twenty. And here we must consider the fact that only those substances which have a specific gravity greater than water can fall through it—substances which must, therefore, be hundreds of times heavier than air; hence when we try to obtain the ratio of the speed in air to that in water, we may, without appreciable error, assume that air does not, to any considerable extent, diminish the free weight [*assoluta gravità*], and consequently the unhindered speed [*assoluta velocità*] of such substances. Having thus easily found the excess of the weight of these substances over that of water, we can say that their speed in air is to their speed in water as their free weight [*totale gravità*] is to the excess of this weight over that of water. For example, a ball of ivory weighs 20 ounces; an equal volume of water weighs 17 ounces; hence the speed of ivory in air bears to its speed in water the approximate ratio of 20:3.

SAGR. I have made a great step forward in this truly interesting subject upon which I have long labored in vain. In order to put these theories into practice we need only discover a method of determining the specific gravity of air with reference to water and hence with reference to other heavy substances.

SIMP. But if we find that air has levity instead of gravity what then shall we say of the foregoing discussion which, in other respects, is very clever?

SALV. I should say that it was empty, vain, and trifling. But can you doubt that air has weight when you have the clear testimony of Aristotle affirming that all the elements have weight including air, and excepting only fire? As evidence of this he cites the fact that a leather bottle weighs more when inflated than when collapsed.

[122]

SIMP. I am inclined to believe that the increase of weight observed in the inflated leather bottle or bladder arises, not from the gravity of the air, but from the many thick vapors mingled with it in these lower regions. To this I would attribute the increase of weight in the leather bottle.

SALV. I would not have you say this, and much less attribute it to Aristotle; because, if speaking of the elements, he wished to persuade

persuade me by experiment that air has weight and were to say to me: "Take a leather bottle, fill it with heavy vapors and observe how its weight increases," I would reply that the bottle would weigh still more if filled with bran; and would then add that this merely proves that bran and thick vapors are heavy, but in regard to air I should still remain in the same doubt as before. However, the experiment of Aristotle is good and the proposition is true. But I cannot say as much of a certain other consideration, taken at face value; this consideration was offered by a philosopher whose name slips me; but I know I have read his argument which is that air exhibits greater gravity than levity, because it carries heavy bodies downward more easily than it does light ones upward.

SAGR. Fine indeed! So according to this theory air is much heavier than water, since all heavy bodies are carried downward more easily through air than through water, and all light bodies buoyed up more easily through water than through air; further there is an infinite number of heavy bodies which fall through air but ascend in water and there is an infinite number of substances which rise in water and fall in air. But, Simplicio, the question as to whether the weight of the leather bottle is owing to thick vapors or to pure air does not affect our problem which is to discover how bodies move through this vapor-laden atmosphere of ours. Returning now to the question which interests me more, I should like, for the sake of more complete and thorough knowledge of this matter, not only to be strengthened in my belief that air has weight but also to learn, if possible, how great its specific gravity is. Therefore, Salviati, if you can satisfy my curiosity on this point pray do so.

SALV. The experiment with the inflated leather bottle of Aristotle proves conclusively that air possesses positive gravity and not, as some have believed, levity, a property possessed possibly by no substance whatever; for if air did possess this quality of absolute and positive levity, it should on compression

[123]

exhibit greater levity and, hence, a greater tendency to rise; but experiment shows precisely the opposite.

As

As to the other question, namely, how to determine the specific gravity of air, I have employed the following method. I took a rather large glass bottle with a narrow neck and attached to it a leather cover, binding it tightly about the neck of the bottle: in the top of this cover I inserted and firmly fastened the valve of a leather bottle, through which I forced into the glass bottle, by means of a syringe, a large quantity of air. And since air is easily condensed one can pump into the bottle two or three times its own volume of air. After this I took an accurate balance and weighed this bottle of compressed air with the utmost precision, adjusting the weight with fine sand. I next opened the valve and allowed the compressed air to escape; then replaced the flask upon the balance and found it perceptibly lighter: from the sand which had been used as a counterweight I now removed and laid aside as much as was necessary to again secure balance. Under these conditions there can be no doubt but that the weight of the sand thus laid aside represents the weight of the air which had been forced into the flask and had afterwards escaped. But after all this experiment tells me merely that the weight of the compressed air is the same as that of the sand removed from the balance; when however it comes to knowing certainly and definitely the weight of air as compared with that of water or any other heavy substance this I cannot hope to do without first measuring the volume [*quantità*] of compressed air; for this measurement I have devised the two following methods.

According to the first method one takes a bottle with a narrow neck similar to the previous one; over the mouth of this bottle is slipped a leather tube which is bound tightly about the neck of the flask; the other end of this tube embraces the valve attached to the first flask and is tightly bound about it. This second flask is provided with a hole in the bottom through which an iron rod can be placed so as to open, at will, the valve above mentioned and thus permit the surplus air of the first to escape after it has once been weighed: but his second bottle must be filled with water. Having prepared everything in the manner

[124]

above

above described, open the valve with the rod; the air will rush
into the flask containing the water and will drive it through the
hole at the bottom, it being clear that the volume [*quantità*] of
water thus displaced is equal to the volume [*mole e quantità*] of
air escaped from the other vessel. Having set aside this displaced water, weigh the vessel from which the air has escaped
(which is supposed to have been weighed previously while
containing the compressed air), and remove the surplus of sand
as described above; it is then manifest that the weight of this
sand is precisely the weight of a volume [*mole*] of air equal to the
volume of water displaced and set aside; this water we can weigh
and find how many times its weight contains the weight of the
removed sand, thus determining definitely how many times
heavier water is than air; and we shall find, contrary to the
opinion of Aristotle, that this is not 10 times, but, as our experiment shows, more nearly 400 times.

The second method is more expeditious and can be carried
out with a single vessel fitted up as the first was. Here no air
is added to that which the vessel naturally contains but water is
forced into it without allowing any air to escape; the water thus
introduced necessarily compresses the air. Having forced into
the vessel as much water as possible, filling it, say, three-fourths
full, which does not require any extraordinary effort, place it
upon the balance and weigh it accurately; next hold the vessel
mouth up, open the valve, and allow the air to escape; the
volume of the air thus escaping is precisely equal to the volume
of water contained in the flask. Again weigh the vessel which
will have diminished in weight on account of the escaped air;
this loss in weight represents the weight of a volume of air equal
to the volume of water contained in the vessel.

Simp. No one can deny the cleverness and ingenuity of your
devices; but while they appear to give complete intellectual
satisfaction they confuse me in another direction. For since it is
undoubtedly true that the elements when in their proper places
have neither weight nor levity, I cannot understand how it is
possible for that portion of air, which appeared to weigh, say,
4 drachms of sand, should really have such a weight in air as the
sand

sand which counterbalances it. It seems to me, therefore, that
the experiment should be carried out, not in air, but in a medium
[125]
in which the air could exhibit its property of weight if such it
really has.

SALV. The objection of Simplicio is certainly to the point and
must therefore either be unanswerable or demand an equally
clear solution. It is perfectly evident that that air which, under
compression, weighed as much as the sand, loses this weight
when once allowed to escape into its own element, while, indeed,
the sand retains its weight. Hence for this experiment it becomes necessary to select a place where air as well as sand can
gravitate; because, as has been often remarked, the medium
diminishes the weight of any substance immersed in it by an
amount equal to the weight of the displaced medium; so that
air in air loses all its weight. If therefore this experiment is to be
made with accuracy it should be performed in a vacuum where
every heavy body exhibits its momentum without the slightest
diminution. If then, Simplicio, we were to weigh a portion of
air in a vacuum would you then be satisfied and assured of the
fact?

SIMP. Yes truly: but this is to wish or ask the impossible.

SALV. Your obligation will then be very great if, for your
sake, I accomplish the impossible. But I do not want to sell you
something which I have already given you; for in the previous
experiment we weighed the air in vacuum and not in air or other
medium. The fact that any fluid medium diminishes the
weight of a mass immersed in it, is due, Simplicio, to the resistance which this medium offers to its being opened up, driven
aside, and finally lifted up. The evidence for this is seen in the
readiness with which the fluid rushes to fill up any space for-
merly occupied by the mass; if the medium were not affected by
such an immersion then it would not react against the immersed
body. Tell me now, when you have a flask, in air, filled with its
natural amount of air and then proceed to pump into the vessel
more air, does this extra charge in any way separate or divide or
change the circumambient air? Does the vessel perhaps expand

so

so that the surrounding medium is displaced in order to give more room? Certainly not. Therefore one is able to say that

[126]

this extra charge of air is not immersed in the surrounding medium for it occupies no space in it, but is, as it were, in a vacuum. Indeed, it is really in a vacuum; for it diffuses into the vacuities which are not completely filled by the original and uncondensed air. In fact I do not see any difference between the enclosed and the surrounding media: for the surrounding medium does not press upon the enclosed medium and, *vice versa*, the enclosed medium exerts no pressure against the surrounding one; this same relationship exists in the case of any matter in a vacuum, as well as in the case of the extra charge of air compressed into the flask. The weight of this condensed air is therefore the same as that which it would have if set free in a vacuum. It is true of course that the weight of the sand used as a counterpoise would be a little greater *in vacuo* than in free air. We must, then, say that the air is slightly lighter than the sand required to counterbalance it, that is to say, by an amount equal to the weight *in vacuo* of a volume of air equal to the volume of the sand.

At this point in an annotated copy of the original edition the following note by Galileo is found.

[Sagr. A very clever discussion, solving a wonderful problem, because it demonstrates briefly and concisely the manner in which one may find the weight of a body *in vacuo* by simply weighing it in air. The explanation is as follows: when a heavy body is immersed in air it loses in weight an amount equal to the weight of a volume [*mole*] of air equivalent to the volume [*mole*] of the body itself. Hence if one adds to a body, without expanding it, a quantity of air equal to that which it displaces and weighs it, he will obtain its absolute weight *in vacuo*, since, without increasing it in size, he has increased its weight by just the amount which it lost through immersion in air.

When therefore we force a quantity of water into a vessel which already contains its normal amount of air, without allowing any of this air to escape it is clear that this normal quantity of air will be compressed and condensed into a smaller space in order to make room for the water which is forced in: it is also clear that the volume of air thus compressed is equal to the volume of water added. If now the vessel be
weighed

weighed in air in this condition, it is manifest that the weight of the water will be increased by that of an equal volume of air; the total weight of water and air thus obtained is equal to the weight of the water alone *in vacuo*.

Now record the weight of the entire vessel and then allow the compressed air to escape; weigh the remainder; the difference of these two weights will be the weight of the compressed air which, in volume, is equal to that of the water. Next find the weight of the water alone and add to it that of the compressed air; we shall then have the water alone *in vacuo*. To find the weight of the water we shall have to remove it from the vessel and weigh the vessel alone; subtraċt this weight from that of the vessel and water together. It is clear that the remainder will be the weight of the water alone in air.]

[127]

SIMP. The previous experiments, in my opinion, left something to be desired: but now I am fully satisfied.

SALV. The faċts set forth by me up to this point and, in particular, the one which shows that difference of weight, even when very great, is without effeċt in changing the speed of falling bodies, so that as far as weight is concerned they all fall with equal speed: this idea is, I say, so new, and at first glance so remote from faċt, that if we do not have the means of making it just as clear as sunlight, it had better not be mentioned; but having once allowed it to pass my lips I must negleċt no experiment or argument to establish it.

SAGR. Not only this but also many other of your views are so far removed from the commonly accepted opinions and doċtrines that if you were to publish them you would stir up a large number of antagonists; for human nature is such that men do not look with favor upon discoveries—either of truth or fallacy—in their own field, when made by others than themselves. They call him an innovator of doċtrine, an unpleasant title, by which they hope to cut those knots which they cannot untie, and by subterranean mines they seek to destroy structures which patient artisans have built with customary tools.

[128]

But as for ourselves who have no such thoughts, the experiments and arguments which you have thus far adduced are fully satisfaċtory; however if you have any experiments which
are

are more direct or any arguments which are more convincing we will hear them with pleasure.

SALV. The experiment made to ascertain whether two bodies, differing greatly in weight will fall from a given height with the same speed offers some difficulty; because, if the height is considerable, the retarding effect of the medium, which must be penetrated and thrust aside by the falling body, will be greater in the case of the small momentum of the very light body than in the case of the great force [*violenza*] of the heavy body; so that, in a long distance, the light body will be left behind; if the height be small, one may well doubt whether there is any difference; and if there be a difference it will be inappreciable.

It occurred to me therefore to repeat many times the fall through a small height in such a way that I might accumulate all those small intervals of time that elapse between the arrival of the heavy and light bodies respectively at their common terminus, so that this sum makes an interval of time which is not only observable, but easily observable. In order to employ the slowest speeds possible and thus reduce the change which the resisting medium produces upon the simple effect of gravity it occurred to me to allow the bodies to fall along a plane slightly inclined to the horizontal. For in such a plane, just as well as in a vertical plane, one may discover how bodies of different weight behave: and besides this, I also wished to rid myself of the resistance which might arise from contact of the moving body with the aforesaid inclined plane. Accordingly I took two balls, one of lead and one of cork, the former more than a hundred times heavier than the latter, and suspended them by means of two equal fine threads, each four or five cubits long. Pulling each ball aside from the perpendicular, I let them go at the same instant, and they, falling along the circumferences of circles having these equal strings for semi-diameters, passed beyond the perpendicular and returned along the same path. This free vibration [*per lor medesime le andate e le tornate*] repeated a hundred times showed clearly that the heavy body maintains so

[129]

nearly the period of the light body that neither in a hundred swings

swings nor even in a thousand will the former anticipate the latter by as much as a single moment [*minimo momento*], so perfectly do they keep step. We can also observe the effect of the medium which, by the resistance which it offers to motion, diminishes the vibration of the cork more than that of the lead, but without altering the frequency of either; even when the arc traversed by the cork did not exceed five or six degrees while that of the lead was fifty or sixty, the swings were performed in equal times.

SIMP. If this be so, why is not the speed of the lead greater than that of the cork, seeing that the former traverses sixty degrees in the same interval in which the latter covers scarcely six?

SALV. But what would you say, Simplicio, if both covered their paths in the same time when the cork, drawn aside through thirty degrees, traverses an arc of sixty, while the lead pulled aside only two degrees traverses an arc of four? Would not then the cork be proportionately swifter? And yet such is the experimental fact. But observe this: having pulled aside the pendulum of lead, say through an arc of fifty degrees, and set it free, it swings beyond the perpendicular almost fifty degrees, thus describing an arc of nearly one hundred degrees; on the return swing it describes a little smaller arc; and after a large number of such vibrations it finally comes to rest. Each vibration, whether of ninety, fifty, twenty, ten, or four degrees occupies the same time: accordingly the speed of the moving body keeps on diminishing since in equal intervals of time, it traverses arcs which grow smaller and smaller.

Precisely the same things happen with the pendulum of cork, suspended by a string of equal length, except that a smaller number of vibrations is required to bring it to rest, since on account of its lightness it is less able to overcome the resistance of the air; nevertheless the vibrations, whether large or small, are all performed in time-intervals which are not only equal among themselves, but also equal to the period of the lead pendulum. Hence it is true that, if while the lead is traversing an arc of fifty degrees the cork covers one of only ten, the cork moves more slowly than the lead; but on the other hand it is also true that

that

[130]

that the cork may cover an arc of fifty while the lead passes over one of only ten or six; thus, at different times, we have now the cork, now the lead, moving more rapidly. But if these same bodies traverse equal arcs in equal times we may rest assured that their speeds are equal.

SIMP. I hesitate to admit the conclusiveness of this argument because of the confusion which arises from your making both bodies move now rapidly, now slowly and now very slowly, which leaves me in doubt as to whether their velocities are always equal.

SAGR. Allow me, if you please, Salviati, to say just a few words. Now tell me, Simplicio, whether you admit that one can say with certainty that the speeds of the cork and the lead are equal whenever both, starting from rest at the same moment and descending the same slopes, always traverse equal spaces in equal times?

SIMP. This can neither be doubted nor gainsaid.

SAGR. Now it happens, in the case of the pendulums, that each of them traverses now an arc of sixty degrees, now one of fifty, or thirty or ten or eight or four or two, etc.; and when they both swing through an arc of sixty degrees they do so in equal intervals of time; the same thing happens when the arc is fifty degrees or thirty or ten or any other number; and therefore we conclude that the speed of the lead in an arc of sixty degrees is equal to the speed of the cork when the latter also swings through an arc of sixty degrees; in the case of a fifty-degree arc these speeds are also equal to each other; so also in the case of other arcs. But this is not saying that the speed which occurs in an arc of sixty is the same as that which occurs in an arc of fifty; nor is the speed in an arc of fifty equal to that in one of thirty, etc.; but the smaller the arcs, the smaller the speeds; the fact observed is that one and the same moving body requires the same time for traversing a large arc of sixty degrees as for a small arc of fifty or even a very small arc of ten; all these arcs, indeed, are covered in the same interval of time. It is true therefore that the lead

[131]

and

and the cork each diminish their speed [*moto*] in proportion as their arcs diminish; but this does not contradict the fact that they maintain equal speeds in equal arcs.

My reason for saying these things has been rather because I wanted to learn whether I had correctly understood Salviati, than because I thought Simplicio had any need of a clearer explanation than that given by Salviati which like everything else of his is extremely lucid, so lucid, indeed, that when he solves questions which are difficult not merely in appearance, but in reality and in fact, he does so with reasons, observations and experiments which are common and familiar to everyone.

In this manner he has, as I have learned from various sources, given occasion to a highly esteemed professor for undervaluing his discoveries on the ground that they are commonplace, and established upon a mean and vulgar basis; as if it were not a most admirable and praiseworthy feature of demonstrative science that it springs from and grows out of principles well-known, understood and conceded by all.

But let us continue with this light diet; and if Simplicio is satisfied to understand and admit that the gravity inherent [*interna gravità*] in various falling bodies has nothing to do with the difference of speed observed among them, and that all bodies, in so far as their speeds depend upon it, would move with the same velocity, pray tell us, Salviati, how you explain the appreciable and evident inequality of motion; please reply also to the objection urged by Simplicio—an objection in which I concur—namely, that a cannon ball falls more rapidly than a bird-shot. From my point of view, one might expect the difference of speed to be small in the case of bodies of the same substance moving through any single medium, whereas the larger ones will descend, during a single pulse-beat, a distance which the smaller ones will not traverse in an hour, or in four, or even in twenty hours; as for instance in the case of stones and fine sand and especially that very fine sand which produces muddy water and which in many hours will not fall through as much as two cubits, a distance which stones not much larger will traverse in a single pulse-beat.

Salv.

SALV. The action of the medium in producing a greater retardation upon those bodies which have a less specific gravity has already been explained by showing that they experience a diminution of weight. But to explain how one and the same

[132]

medium produces such different retardations in bodies which are made of the same material and have the same shape, but differ only in size, requires a discussion more clever than that by which one explains how a more expanded shape or an opposing motion of the medium retards the speed of the moving body. The solution of the present problem lies, I think, in the roughness and porosity which are generally and almost necessarily found in the surfaces of solid bodies. When the body is in motion these rough places strike the air or other ambient medium. The evidence for this is found in the humming which accompanies the rapid motion of a body through air, even when that body is as round as possible. One hears not only humming, but also hissing and whistling, whenever there is any appreciable cavity or elevation upon the body. We observe also that a round solid body rotating in a lathe produces a current of air. But what more do we need? When a top spins on the ground at its greatest speed do we not hear a distinct buzzing of high pitch? This sibilant note diminishes in pitch as the speed of rotation slackens, which is evidence that these small rugosities on the surface meet resistance in the air. There can be no doubt, therefore, that in the motion of falling bodies these rugosities strike the surrounding fluid and retard the speed; and this they do so much the more in proportion as the surface is larger, which is the case of small bodies as compared with greater.

SIMP. Stop a moment please, I am getting confused. For although I understand and admit that friction of the medium upon the surface of the body retards its motion and that, if other things are the same, the larger surface suffers greater retardation, I do not see on what ground you say that the surface of the smaller body is larger. Besides if, as you say, the larger surface suffers greater retardation the larger solid should move more slowly, which is not the fact. But this objection can be easily met

met by saying that, although the larger body has a larger surface, it has also a greater weight, in comparison with which the resistance of the larger surface is no more than the resistance of the small surface in comparison with its smaller weight; so that the speed of the larger solid does not become less. I therefore see no reason for expecting any difference of speed so long as the driving weight [*gravità movente*] diminishes in the same propor-

[133]

tion as the retarding power [*facoltà ritardante*] of the surface.

SALV. I shall answer all your objections at once. You will admit, of course, Simplicio, that if one takes two equal bodies, of the same material and same figure, bodies which would therefore fall with equal speeds, and if he diminishes the weight of one of them in the same proportion as its surface (maintaining the similarity of shape) he would not thereby diminish the speed of this body.

SIMP. This inference seems to be in harmony with your theory which states that the weight of a body has no effect in either accelerating or retarding its motion.

SALV. I quite agree with you in this opinion from which it appears to follow that, if the weight of a body is diminished in greater proportion than its surface, the motion is retarded to a certain extent; and this retardation is greater and greater in proportion as the diminution of weight exceeds that of the surface.

SIMP. This I admit without hesitation.

SALV. Now you must know, Simplicio, that it is not possible to diminish the surface of a solid body in the same ratio as the weight, and at the same time maintain similarity of figure. For since it is clear that in the case of a diminishing solid the weight grows less in proportion to the volume, and since the volume always diminishes more rapidly than the surface, when the same shape is maintained, the weight must therefore diminish more rapidly than the surface. But geometry teaches us that, in the case of similar solids, the ratio of two volumes is greater than the ratio of their surfaces; which, for the sake of better understanding, I shall illustrate by a particular case.

Take,

Take, for example, a cube two inches on a side so that each face has an area of four square inches and the total area, i. e., the sum of the six faces, amounts to twenty-four square inches; now imagine this cube to be sawed through three times so as to divide it into eight smaller cubes, each one inch on the side, each face one inch square, and the total surface of each cube six square inches instead of twenty-four as in the case of the larger

[134]

cube. It is evident therefore that the surface of the little cube is only one-fourth that of the larger, namely, the ratio of six to twenty-four; but the volume of the solid cube itself is only one-eighth; the volume, and hence also the weight, diminishes therefore much more rapidly than the surface. If we again divide the little cube into eight others we shall have, for the total surface of one of these, one and one-half square inches, which is one-sixteenth of the surface of the original cube; but its volume is only one-sixty-fourth part. Thus, by two divisions, you see that the volume is diminished four times as much as the surface. And, if the subdivision be continued until the original solid be reduced to a fine powder, we shall find that the weight of one of these smallest particles has diminished hundreds and hundreds of times as much as its surface. And this which I have illustrated in the case of cubes holds also in the case of all similar solids, where the volumes stand in sesquialteral ratio to their surfaces. Observe then how much greater the resistance, arising from contact of the surface of the moving body with the medium, in the case of small bodies than in the case of large; and when one considers that the rugosities on the very small surfaces of fine dust particles are perhaps no smaller than those on the surfaces of larger solids which have been carefully polished, he will see how important it is that the medium should be very fluid and offer no resistance to being thrust aside, easily yielding to a small force. You see, therefore, Simplicio, that I was not mistaken when, not long ago, I said that the surface of a small solid is comparatively greater than that of a large one.

SIMP. I am quite convinced; and, believe me, if I were again beginning my studies, I should follow the advice of Plato and start

start with mathematics, a science which proceeds very cautiously and admits nothing as established until it has been rigidly demonstrated.

SAGR. This discussion has afforded me great pleasure; but before proceeding further I should like to hear the explanation of a phrase of yours which is new to me, namely, that similar solids are to each other in the sesquialteral ratio of their surfaces; for although I have seen and understood the proposition in which it is demonstrated that the surfaces of similar solids are in the
[135]
duplicate ratio of their sides and also the proposition which proves that the volumes are in the triplicate ratio of their sides, yet I have not so much as heard mentioned the ratio of the volume of a solid to its surface.

SALV. You yourself have suggested the answer to your question and have removed every doubt. For if one quantity is the cube of something of which another quantity is the square does it not follow that the cube is the sesquialteral of the square? Surely. Now if the surface varies as the square of its linear dimensions while the volume varies as the cube of these dimensions may we not say that the volume stands in sesquialteral ratio to the surface?

SAGR. Quite so. And now although there are still some details, in connection with the subject under discussion, concerning which I might ask questions yet, if we keep making one digression after another, it will be long before we reach the main topic which has to do with the variety of properties found in the resistance which solid bodies offer to fracture; and, therefore, if you please, let us return to the subject which we originally proposed to discuss.

SALV. Very well; but the questions which we have already considered are so numerous and so varied, and have taken up so much time that there is not much of this day left to spend upon our main topic which abounds in geometrical demonstrations calling for careful consideration. May I, therefore, suggest that we postpone the meeting until to-morrow, not only for the reason just mentioned but also in order that I may bring with me me some papers in which I have set down in an orderly way the theorems and propositions dealing with the various phases of this subject, matters which, from memory alone, I could not present in the proper order.

SAGR. I fully concur in your opinion and all the more willingly because this will leave time to-day to take up some of my difficulties with the subject which we have just been discussing. One question is whether we are to consider the resistance of the medium as sufficient to destroy the acceleration of a body of very heavy material, very large volume, and
[136]
spherical figure. I say *spherical* in order to select a volume which is contained within a minimum surface and therefore less subject to retardation.

Another question deals with the vibrations of pendulums which may be regarded from several viewpoints; the first is whether all vibrations, large, medium, and small, are performed in exactly and precisely equal times: another is to find the ratio of the times of vibration of pendulums supported by threads of unequal length.

SALV. These are interesting questions: but I fear that here, as in the case of all other facts, if we take up for discussion any one of them, it will carry in its wake so many other facts and curious consequences that time will not remain to-day for the discussion of all.

SAGR. If these are as full of interest as the foregoing, I would gladly spend as many days as there remain hours between now and nightfall; and I dare say that Simplicio would not be wearied by these discussions.

SIMP. Certainly not; especially when the questions pertain to natural science and have not been treated by other philosophers.

SALV. Now taking up the first question, I can assert without hesitation that there is no sphere so large, or composed of material so dense but that the resistance of the medium, although very slight, would check its acceleration and would, in time reduce its motion to uniformity; a statement which is
strongly

strongly supported by experiment. For if a falling body, as time goes on, were to acquire a speed as great as you please, no such speed, impressed by external forces [*motore esterno*], can be so great but that the body will first acquire it and then, owing to the resisting medium, lose it. Thus, for instance, if a cannon ball, having fallen a distance of four cubits through the air and having acquired a speed of, say, ten units [*gradi*] were to strike the surface of the water, and if the resistance of the water were not able to check the momentum [*impeto*] of the shot, it would either increase in speed or maintain a uniform motion until the bottom were reached: but such is not the observed fact; on the contrary, the water when only a few cubits deep hinders and diminishes the motion in such a way that the shot delivers to the bed of the river or lake a very slight impulse. Clearly

[137]

then if a short fall through the water is sufficient to deprive a cannon ball of its speed, this speed cannot be regained by a fall of even a thousand cubits. How could a body acquire, in a fall of a thousand cubits, that which it loses in a fall of four? But what more is needed? Do we not observe that the enormous momentum, delivered to a shot by a cannon, is so deadened by passing through a few cubits of water that the ball, so far from injuring the ship, barely strikes it? Even the air, although a very yielding medium, can also diminish the speed of a falling body, as may be easily understood from similar experiments. For if a gun be fired downwards from the top of a very high tower the shot will make a smaller impression upon the ground than if the gun had been fired from an elevation of only four or six cubits; this is clear evidence that the momentum of the ball, fired from the top of the tower, diminishes continually from the instant it leaves the barrel until it reaches the ground. Therefore a fall from ever so great an altitude will not suffice to give to a body that momentum which it has once lost through the resistance of the air, no matter how it was originally acquired. In like manner, the destructive effect produced upon a wall by a shot fired from a gun at a distance of twenty cubits cannot be duplicated by the fall of the same shot from any altitude how-

ever

ever great. My opinion is, therefore, that under the circumstances which occur in nature, the acceleration of any body falling from rest reaches an end and that the resistance of the medium finally reduces its speed to a constant value which is thereafter maintained.

SAGR. These experiments are in my opinion much to the purpose; the only question is whether an opponent might not make bold to deny the fact in the case of bodies [*moli*] which are very large and heavy or to assert that a cannon ball, falling from the distance of the moon or from the upper regions of the atmosphere, would deliver a heavier blow than if just leaving the muzzle of the gun.

SALV. No doubt many objections may be raised not all of which can be refuted by experiment: however in this particular

[138]

case the following consideration must be taken into account, namely, that it is very likely that a heavy body falling from a height will, on reaching the ground, have acquired just as much momentum as was necessary to carry it to that height; as may be clearly seen in the case of a rather heavy pendulum which, when pulled aside fifty or sixty degrees from the vertical, will acquire precisely that speed and force which are sufficient to carry it to an equal elevation save only that small portion which it loses through friction on the air. In order to place a cannon ball at such a height as might suffice to give it just that momentum which the powder imparted to it on leaving the gun we need only fire it vertically upwards from the same gun; and we can then observe whether on falling back it delivers a blow equal to that of the gun fired at close range; in my opinion it would be much weaker. The resistance of the air would, therefore, I think, prevent the muzzle velocity from being equalled by a natural fall from rest at any height whatsoever.

We come now to the other questions, relating to pendulums, a subject which may appear to many exceedingly arid, especially to those philosophers who are continually occupied with the more profound questions of nature. Nevertheless, the problem is one which I do not scorn. I am encouraged by the

example

example of Aristotle whom I admire especially because he did
not fail to discuss every subject which he thought in any degree
worthy of consideration.

Impelled by your queries I may give you some of my ideas
concerning certain problems in music, a splendid subject, upon
which so many eminent men have written: among these is
Aristotle himself who has discussed numerous interesting acoustical questions. Accordingly, if on the basis of some easy and
tangible experiments, I shall explain some striking phenomena
in the domain of sound, I trust my explanations will meet your
approval.

Sagr. I shall receive them not only gratefully but eagerly.
For, although I take pleasure in every kind of musical instru-

[139]

ment and have paid considerable attention to harmony, I have
never been able to fully understand why some combinations of
tones are more pleasing than others, or why certain combinations not only fail to please but are even highly offensive.
Then there is the old problem of two stretched strings in unison;
when one of them is sounded, the other begins to vibrate and
to emit its note; nor do I understand the different ratios of
harmony [*forme delle consonanze*] and some other details.

Salv. Let us see whether we cannot derive from the pendulum
a satisfactory solution of all these difficulties. And first, as to
the question whether one and the same pendulum really performs its vibrations, large, medium, and small, all in exactly
the same time, I shall rely upon what I have already heard from
our Academician. He has clearly shown that the time of
descent is the same along all chords, whatever the arcs which
subtend them, as well along an arc of 180° (i. e., the whole
diameter) as along one of 100°, 60°, 10°, 2°, ½°, or 4′. It is
understood, of course, that these arcs all terminate at the
lowest point of the circle, where it touches the horizontal plane.

If now we consider descent along arcs instead of their chords
then, provided these do not exceed 90°, experiment shows that
they are all traversed in equal times; but these times are greater
for the chord than for the arc, an effect which is all the more
                                                    remarkable

remarkable because at first glance one would think just the opposite to be true. For since the terminal points of the two motions are the same and since the straight line included between these two points is the shortest distance between them, it would seem reasonable that motion along this line should be executed in the shortest time; but this is not the case, for the shortest time—and therefore the most rapid motion—is that employed along the arc of which this straight line is the chord.

As to the times of vibration of bodies suspended by threads of different lengths, they bear to each other the same proportion as the square roots of the lengths of the thread; or one might say the lengths are to each other as the squares of the times; so that if one wishes to make the vibration-time of one pendulum twice that of another, he must make its suspension four times as long. In like manner, if one pendulum has a suspension nine times as

[140]

long as another, this second pendulum will execute three vibrations during each one of the first; from which it follows that the lengths of the suspending cords bear to each other the [inverse] ratio of the squares of the number of vibrations performed in the same time.

SAGR. Then, if I understand you correctly, I can easily measure the length of a string whose upper end is attached at any height whatever even if this end were invisible and I could see only the lower extremity. For if I attach to the lower end of this string a rather heavy weight and give it a to-and-fro motion, and if I ask a friend to count a number of its vibrations, while I, during the same time-interval, count the number of vibrations of a pendulum which is exactly one cubit in length, then knowing the number of vibrations which each pendulum makes in the given interval of time one can determine the length of the string. Suppose, for example, that my friend counts 20 vibrations of the long cord during the same time in which I count 240 of my string which is one cubit in length; taking the squares of the two numbers, 20 and 240, namely 400 and 57600, then, I say, the long string contains 57600 units of such length that my pendulum will contain 400 of them; and since the length of

my

my string is one cubit, I shall divide 57600 by 400 and thus obtain 144. Accordingly I shall call the length of the string 144 cubits.

SALV. Nor will you miss it by as much as a hand's breadth, especially if you observe a large number of vibrations.

SAGR. You give me frequent occasion to admire the wealth and profusion of nature when, from such common and even trivial phenomena, you derive facts which are not only striking and new but which are often far removed from what we would have imagined. Thousands of times I have observed vibrations especially in churches where lamps, suspended by long cords, had been inadvertently set into motion; but the most which I could infer from these observations was that the view of those who think that such vibrations are maintained by the medium is highly improbable: for, in that case, the air must needs have considerable judgment and little else to do but kill time by pushing to and fro a pendent weight with perfect regularity. But I never dreamed of learning that one and the same body, when
[141]
suspended from a string a hundred cubits long and pulled aside through an arc of 90° or even 1° or ½°, would employ the same time in passing through the least as through the largest of these arcs; and, indeed, it still strikes me as somewhat unlikely. Now I am waiting to hear how these same simple phenomena can furnish solutions for those acoustical problems—solutions which will be at least partly satisfactory.

SALV. First of all one must observe that each pendulum has its own time of vibration so definite and determinate that it is not possible to make it move with any other period [*altro periodo*] than that which nature has given it. For let any one take in his hand the cord to which the weight is attached and try, as much as he pleases, to increase or diminish the frequency [*frequenza*] of its vibrations; it will be time wasted. On the other hand, one can confer motion upon even a heavy pendulum which is at rest by simply blowing against it; by repeating these blasts with a frequency which is the same as that of the pendulum one can impart considerable motion. Suppose that by the
first

first puff we have displaced the pendulum from the vertical by, say, half an inch; then if, after the pendulum has returned and is about to begin the second vibration, we add a second puff, we shall impart additional motion; and so on with other blasts provided they are applied at the right instant, and not when the pendulum is coming toward us since in this case the blast would impede rather than aid the motion. Continuing thus with many impulses [*impulsi*] we impart to the pendulum such momentum [*impeto*] that a greater impulse [*forza*] than that of a single blast will be needed to stop it.

SAGR. Even as a boy, I observed that one man alone by giving these impulses at the right instant was able to ring a bell so large that when four, or even six, men seized the rope and tried to stop it they were lifted from the ground, all of them together being unable to counterbalance the momentum which a single man, by properly-timed pulls, had given it.

SALV. Your illustration makes my meaning clear and is quite as well fitted, as what I have just said, to explain the wonderful phenomenon of the strings of the cittern [*cetera*] or of the spinet

[142]

[*cimbalo*], namely, the fact that a vibrating string will set another string in motion and cause it to sound not only when the latter is in unison but even when it differs from the former by an octave or a fifth. A string which has been struck begins to vibrate and continues the motion as long as one hears the sound [*risonanza*]; these vibrations cause the immediately surrounding air to vibrate and quiver; then these ripples in the air expand far into space and strike not only all the strings of the same instrument but even those of neighboring instruments. Since that string which is tuned to unison with the one plucked is capable of vibrating with the same frequency, it acquires, at the first impulse, a slight oscillation; after receiving two, three, twenty, or more impulses, delivered at proper intervals, it finally accumulates a vibratory motion equal to that of the plucked string, as is clearly shown by equality of amplitude in their vibrations. This undulation expands through the air and sets into vibration not only strings, but also any other body which

which happens to have the same period as that of the plucked string. Accordingly if we attach to the side of an instrument small pieces of bristle or other flexible bodies, we shall observe that, when a spinet is sounded, only those pieces respond that have the same period as the string which has been struck; the remaining pieces do not vibrate in response to this string, nor do the former pieces respond to any other tone.

If one bows the base string on a viola rather smartly and brings near it a goblet of fine, thin glass having the same tone [*tuono*] as that of the string, this goblet will vibrate and audibly resound. That the undulations of the medium are widely dispersed about the sounding body is evinced by the fact that a glass of water may be made to emit a tone merely by the friction of the finger-tip upon the rim of the glass; for in this water is produced a series of regular waves. The same phenomenon is observed to better advantage by fixing the base of the goblet upon the bottom of a rather large vessel of water filled nearly to the edge of the goblet; for if, as before, we sound the glass by friction of the finger, we shall see ripples spreading with the utmost regularity and with high speed to large distances about the glass. I have often remarked, in thus sounding a rather

[143]

large glass nearly full of water, that at first the waves are spaced with great uniformity, and when, as sometimes happens, the tone of the glass jumps an octave higher I have noted that at this moment each of the aforesaid waves divides into two; a phenomenon which shows clearly that the ratio involved in the octave [*forma dell' ottava*] is two.

SAGR. More than once have I observed this same thing, much to my delight and also to my profit. For a long time I have been perplexed about these different harmonies since the explanations hitherto given by those learned in music impress me as not sufficiently conclusive. They tell us that the diapason, i. e. the octave, involves the ratio of two, that the diapente which we call the fifth involves a ratio of 3:2, etc.; because if the open string of a monochord be sounded and afterwards a bridge be placed in the middle and the half length be sounded

one

one hears the octave; and if the bridge be placed at 1/3 the length of the string, then on plucking first the open string and afterwards 2/3 of its length the fifth is given; for this reason they say that the octave depends upon the ratio of two to one [*contenuta tra'l due e l'uno*] and the fifth upon the ratio of three to two. This explanation does not impress me as sufficient to establish 2 and 3/2 as the natural ratios of the octave and the fifth; and my reason for thinking so is as follows. There are three different ways in which the tone of a string may be sharpened, namely, by shortening it, by stretching it and by making it thinner. If the tension and size of the string remain constant one obtains the octave by shortening it to one-half, i. e., by sounding first the open string and then one-half of it; but if length and size remain constant and one attempts to produce the octave by stretching he will find that it does not suffice to double the stretching weight; it must be quadrupled; so that, if the fundamental note is produced by a weight of one pound, four will be required to bring out the octave.

And finally if the length and tension remain constant, while one changes the size * of the string he will find that in order to produce the octave the size must be reduced to 1/4 that which gave the fundamental. And what I have said concerning the octave, namely, that its ratio as derived from the tension and size of the string is the square of that derived from the length, applies equally well to all other musical intervals [*intervalli*
[144]
musici]. Thus if one wishes to produce a fifth by changing the length he finds that the ratio of the lengths must be sesquialteral, in other words he sounds first the open string, then two-thirds of it; but if he wishes to produce this same result by stretching or thinning the string then it becomes necessary to square the ratio 3/2 that is by taking 9/4 [*dupla sesquiquarta*]; accordingly, if the fundamental requires a weight of 4 pounds, the higher note will be produced not by 6, but by 9 pounds; the same is true in regard to size, the string which gives the fundamental is larger than that which yields the fifth in the ratio of 9 to 4.

In view of these facts, I see no reason why those wise philos-

* For the exact meaning of "size" see p. 103 below. [*Trans.*]

ophers should adopt 2 rather than 4 as the ratio of the octave, or why in the case of the fifth they should employ the sesquialteral ratio, 3/2, rather than that of 9/4. Since it is impossible to count the vibrations of a sounding string on account of its high frequency, I should still have been in doubt as to whether a string, emitting the upper octave, made twice as many vibrations in the same time as one giving the fundamental, had it not been for the following fact, namely, that at the instant when the tone jumps to the octave, the waves which constantly accompany the vibrating glass divide up into smaller ones which are precisely half as long as the former.

SALV. This is a beautiful experiment enabling us to distinguish individually the waves which are produced by the vibrations of a sonorous body, which spread through the air, bringing to the tympanum of the ear a stimulus which the mind translates into sound. But since these waves in the water last only so long as the friction of the finger continues and are, even then, not constant but are always forming and disappearing, would it not be a fine thing if one had the ability to produce waves which would persist for a long while, even months and years, so as to easily measure and count them?

SAGR. Such an invention would, I assure you, command my admiration.

SALV. The device is one which I hit upon by accident; my part consists merely in the observation of it and in the appreciation of its value as a confirmation of something to which I had given profound consideration; and yet the device is, in itself, rather common. As I was scraping a brass plate with a sharp iron
[145]
chisel in order to remove some spots from it and was running the chisel rather rapidly over it, I once or twice, during many strokes, heard the plate emit a rather strong and clear whistling sound; on looking at the plate more carefully, I noticed a long row of fine streaks parallel and equidistant from one another. Scraping with the chisel over and over again, I noticed that it was only when the plate emitted this hissing noise that any marks were left upon it; when the scraping was not accompanied
by

by this sibilant note there was not the least trace of such marks. Repeating the trick several times and making the stroke, now with greater now with less speed, the whistling followed with a pitch which was correspondingly higher and lower. I noted also that the marks made when the tones were higher were closer together; but when the tones were deeper, they were farther apart. I also observed that when, during a single stroke, the speed increased toward the end the sound became sharper and the streaks grew closer together, but always in such a way as to remain sharply defined and equidistant. Besides whenever the stroke was accompanied by hissing I felt the chisel tremble in my grasp and a sort of shiver run through my hand. In short we see and hear in the case of the chisel precisely that which is seen and heard in the case of a whisper followed by a loud voice; for, when the breath is emitted without the production of a tone, one does not feel either in the throat or mouth any motion to speak of in comparison with that which is felt in the larynx and upper part of the throat when the voice is used, especially when the tones employed are low and strong.

At times I have also observed among the strings of the spinet two which were in unison with two of the tones produced by the aforesaid scraping; and among those which differed most in pitch I found two which were separated by an interval of a perfect fifth. Upon measuring the distance between the markings produced by the two scrapings it was found that the space which contained 45 of one contained 30 of the other, which is precisely the ratio assigned to the fifth.

But now before proceeding any farther I want to call your attention to the fact that, of the three methods for sharpening a tone, the one which you refer to as the fineness of the string should be attributed to its weight. So long as the material of
[146]
the string is unchanged, the size and weight vary in the same ratio. Thus in the case of gut-strings, we obtain the octave by making one string 4 times as large as the other; so also in the case of brass one wire must have 4 times the size of the other; but if now we wish to obtain the octave of a gut-string, by use of
brass

brass wire, we must make it, not four times as large, but four times as heavy as the gut-string: as regards size therefore the metal string is not four times as big but four times as heavy. The wire may therefore be even thinner than the gut notwithstanding the fact that the latter gives the higher note. Hence if two spinets are strung, one with gold wire the other with brass, and if the corresponding strings each have the same length, diameter, and tension it follows that the instrument strung with gold will have a pitch about one-fifth lower than the other because gold has a density almost twice that of brass. And here it is to be noted that it is the weight rather than the size of a moving body which offers resistance to change of motion [*velocità del moto*] contrary to what one might at first glance think. For it seems reasonable to believe that a body which is large and light should suffer greater retardation of motion in thrusting aside the medium than would one which is thin and heavy; yet here exactly the opposite is true.

Returning now to the original subject of discussion, I assert that the ratio of a musical interval is not immediately determined either by the length, size, or tension of the strings but rather by the ratio of their frequencies, that is, by the number of pulses of air waves which strike the tympanum of the ear, causing it also to vibrate with the same frequency. This fact established, we may possibly explain why certain pairs of notes, differing in pitch produce a pleasing sensation, others a less pleasant effect, and still others a disagreeable sensation. Such an explanation would be tantamount to an explanation of the more or less perfect consonances and of dissonances. The unpleasant sensation produced by the latter arises, I think, from the discordant vibrations of two different tones which strike the ear out of time [*sproporzionatamente*]. Especially harsh is the dissonance between notes whose frequencies are incommensurable; such a case occurs when one has two strings in unison and sounds one of them open, together with a part of the other

[147]

which bears the same ratio to its wnole length as the side of a square bears to the diagonal; this yields a dissonance similar

to

to the augmented fourth or diminished fifth [*tritono o semi-diapente*].

Agreeable consonances are pairs of tones which strike the ear with a certain regularity; this regularity consists in the fact that the pulses delivered by the two tones, in the same interval of time, shall be commensurable in number, so as not to keep the ear drum in perpetual torment, bending in two different directions in order to yield to the ever-discordant impulses.

The first and most pleasing consonance is, therefore, the octave since, for every pulse given to the tympanum by the lower string, the sharp string delivers two; accordingly at every other vibration of the upper string both pulses are delivered simultaneously so that one-half the entire number of pulses are delivered in unison. But when two strings are in unison their vibrations always coincide and the effect is that of a single string; hence we do not refer to it as consonance. The fifth is also a pleasing interval since for every two vibrations of the lower string the upper one gives three, so that considering the entire number of pulses from the upper string one-third of them will strike in unison, i. e., between each pair of concordant vibrations there intervene two single vibrations; and when the interval is a fourth, three single vibrations intervene. In case the interval is a second where the ratio is 9/8 it is only every ninth vibration of the upper string which reaches the ear simultaneously with one of the lower; all the others are discordant and produce a harsh effect upon the recipient ear which interprets them as dissonances.

SIMP. Won't you be good enough to explain this argument a little more clearly?

SALV. Let AB denote the length of a wave [*lo spazio e la dilatazione d'una vibrazione*] emitted by the lower string and CD that of a higher string which is emitting the octave of AB; divide AB in the middle at E. If the two strings begin their motions at A and C, it is clear that when the sharp vibration has reached the end D, the other vibration will have travelled only as far as E, which, not being a terminal point, will emit no pulse; but there is a blow delivered at D. Accordingly when the one

wave

wave comes back from D to C, the other passes on from E to B; hence the two pulses from B and C strike the drum of the ear simultaneously. Seeing that these vibrations are repeated again and again in the same manner, we conclude that each alternate pulse from CD falls in unison with one from AB. But each of the [148] pulsations at the terminal points, A and B, is constantly accompanied by one which leaves always from C or always from D. This is clear because if we suppose the waves to reach A and C at the same instant, then, while one wave travels from A to B, the other will proceed from C to D and back to C, so that waves strike at C and B simultaneously; during the passage of the wave from B back to A the disturbance at C goes to D and again returns to C, so that once more the pulses at A and C are simultaneous.

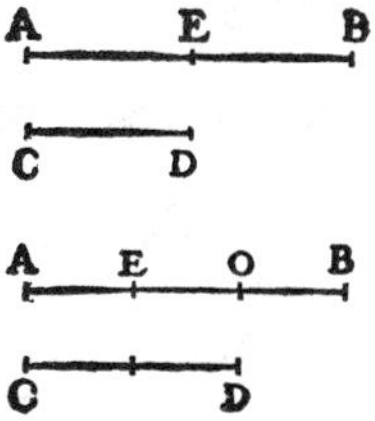

Fig. 13

Next let the vibrations AB and CD be separated by an interval of a fifth, that is, by a ratio of $3/2$; choose the points E and O such that they will divide the wave length of the lower string into three equal parts and imagine the vibrations to start at the same instant from each of the terminals A and C. It is evident that when the pulse has been delivered at the terminal D, the wave in AB has travelled only as far as O; the drum of the ear receives, therefore, only the pulse from D. Then during the return of the one vibration from D to C, the other will pass from O to B and then back to O, producing an isolated pulse at B—a pulse which is out of time but one which must be taken into consideration.

Now since we have assumed that the first pulsations started from the terminals A and C at the same instant, it follows that the second pulsation, isolated at D, occurred after an interval of time equal to that required for passage from C to D or, what is the same thing, from A to O; but the next pulsation, the one at B, is separated from the preceding by only half this interval, namely, the time required for passage from O to B. Next while the one vibration travels from O to A, the other travels from C to D,

D, the result of which is that two pulsations occur simultaneously at A and D. Cycles of this kind follow one after another, i. e., one solitary pulse of the lower string interposed between two solitary pulses of the upper string. Let us now imagine time to be divided into very small equal intervals; then if we assume that, during the first two of these intervals, the disturbances which occurred simultaneously at A and C have travelled as far as O and D and have produced a pulse at D; and if we assume that during the third and fourth intervals one disturbance returns from D to C, producing a pulse at C, while the other, passing on from O to B and back to O, produces a pulse at B; and if finally, during the fifth and sixth intervals, the disturbances travel from O and C to A and D, producing a pulse at each of the latter two, then the sequence in which the pulses strike the ear will be such that, if we begin to count time from any instant where two pulses are simultaneous, the ear drum will, after the lapse of two of the said intervals, receive a solitary pulse; at the end of the third interval, another solitary

[149]

pulse; so also at the end of the fourth interval; and two intervals later, i. e., at the end of the sixth interval, will be heard two pulses in unison. Here ends the cycle—the anomaly, so to speak—which repeats itself over and over again.

SAGR. I can no longer remain silent; for I must express to you the great pleasure I have in hearing such a complete explanation of phenomena with regard to which I have so long been in darkness. Now I understand why unison does not differ from a single tone; I understand why the octave is the principal harmony, but so like unison as often to be mistaken for it and also why it occurs with the other harmonies. It resembles unison because the pulsations of strings in unison always occur simultaneously, and those of the lower string of the octave are always accompanied by those of the upper string; and among the latter is interposed a solitary pulse at equal intervals and in such a manner as to produce no disturbance; the result is that such a harmony is rather too much softened and lacks fire. But the fifth is characterized by its displaced beats and by the interposition

tion of two solitary beats of the upper string and one solitary beat of the lower string between each pair of simultaneous pulses; these three solitary pulses are separated by intervals of time equal to half the interval which separates each pair of simultaneous beats from the solitary beats of the upper string. Thus the effect of the fifth is to produce a tickling of the ear drum such that its softness is modified with sprightliness, giving at the same moment the impression of a gentle kiss and of a bite.

SALV. Seeing that you have derived so much pleasure from these novelties, I must show you a method by which the eye may enjoy the same game as the ear. Suspend three balls of lead, or other heavy material, by means of strings of different length such that while the longest makes two vibrations the shortest will make four and the medium three; this will take place when the longest string measures 16, either in hand breadths or in any other unit, the medium 9 and the shortest 4, all measured in the same unit.

Now pull all these pendulums aside from the perpendicular and release them at the same instant; you will see a curious interplay of the threads passing each other in various manners but such that at the completion of every fourth vibration of the longest pendulum, all three will arrive simultaneously at the same terminus, whence they start over again to repeat the same cycle. This combination of vibrations, when produced on strings is precisely that which yields the interval of the octave and the intermediate fifth. If we employ the same disposition

[150]

of apparatus but change the lengths of the threads, always however in such a way that their vibrations correspond to those of agreeable musical intervals, we shall see a different crossing of these threads but always such that, after a definite interval of time and after a definite number of vibrations, all the threads, whether three or four, will reach the same terminus at the same instant, and then begin a repetition of the cycle.

If however the vibrations of two or more strings are incommensurable so that they never complete a definite number of vibrations at the same instant, or if commensurable they return only

only after a long interval of time and after a large number of vibrations, then the eye is confused by the disorderly succession of crossed threads. In like manner the ear is pained by an irregular sequence of air waves which strike the tympanum without any fixed order.

But, gentlemen, whither have we drifted during these many hours lured on by various problems and unexpected digressions? The day is already ended and we have scarcely touched the subject proposed for discussion. Indeed we have deviated so far that I remember only with difficulty our early introduction and the little progress made in the way of hypotheses and principles for use in later demonstrations.

SAGR. Let us then adjourn for to-day in order that our minds may find refreshment in sleep and that we may return to-morrow, if so please you, and resume the discussion of the main question.

SALV. I shall not fail to be here to-morrow at the same hour, hoping not only to render you service but also to enjoy your company.

END OF THE FIRST DAY.

SECOND DAY

AGR. While Simplicio and I were awaiting your arrival we were trying to recall that last consideration which you advanced as a principle and basis for the results you intended to obtain; this consideration dealt with the resistance which all solids offer to fracture and depended upon a certain cement which held the parts glued together so that they would yield and separate only under considerable pull [*potente attrazzione*]. Later we tried to find the explanation of this coherence, seeking it mainly in the vacuum; this was the occasion of our many digressions which occupied the entire day and led us far afield from the original question which, as I have already stated, was the consideration of the resistance [*resistenza*] that solids offer to fracture.

SALV. I remember it all very well. Resuming the thread of our discourse, whatever the nature of this resistance which solids offer to large tractive forces [*violenta attrazzione*] there can at least be no doubt of its existence; and though this resistance is very great in the case of a direct pull, it is found, as a rule, to be less in the case of bending forces [*nel violentargli per traverso*]. Thus, for example, a rod of steel or of glass will sustain a longitudinal pull of a thousand pounds while a weight of fifty pounds would be quite sufficient to break it if the rod were fastened at right angles into a vertical wall. It is this second type of resistance which we must consider, seeking to discover in what

proportion

proportion it is found in prisms and cylinders of the same material, whether alike or unlike in shape, length, and thickness. In this discussion I shall take for granted the well-known mechanical principle which has been shown to govern the behavior of a bar, which we call a lever, namely, that the force bears to the resistance the inverse ratio of the distances which separate the fulcrum from the force and resistance respectively.

SIMP. This was demonstrated first of all by Aristotle, in his *Mechanics.*

SALV. Yes, I am willing to concede him priority in point of time; but as regards rigor of demonstration the first place must be given to Archimedes, since upon a single proposition proved in his book on *Equilibrium* * depends not only the law of the lever but also those of most other mechanical devices.

SAGR. Since now this principle is fundamental to all the demonstrations which you propose to set forth would it not be advisable to give us a complete and thorough proof of this proposition unless possibly it would take too much time?

SALV. Yes, that would be quite proper, but it is better I think to approach our subject in a manner somewhat different from that employed by Archimedes, namely, by first assuming merely that equal weights placed in a balance of equal arms will produce equilibrium—a principle also assumed by Archimedes—and then proving that it is no less true that unequal weights produce equilibrium when the arms of the steelyard have lengths inversely proportional to the weights suspended from them; in other words, it amounts to the same thing whether one places equal weights at equal distances or unequal weights at distances which bear to each other the inverse ratio of the weights.

In order to make this matter clear imagine a prism or solid cylinder, AB, suspended at each end to the rod [*linea*] HI, and supported by two threads HA and IB; it is evident that if I attach a thread, C, at the middle point of the balance beam HI, the entire prism AB will, according to the principle assumed, hang in equilibrium since one-half its weight lies on one side, and the other half on the other side, of the point of suspension C. Now

* *Works of Archimedes.* Trans. by T. L. Heath, pp. 189–220. [*Trans.*]

suppose the prism to be divided into unequal parts by a plane
[153]
through the line D, and let the part DA be the larger and DB
the smaller: this division having been made, imagine a thread
ED, attached at the point E and supporting the parts AD and
DB, in order that these parts may remain in the same position
relative to line HI: and since the relative position of the prism
and the beam HI remains unchanged, there can be no doubt
but that the prism will maintain its former state of equilibrium.

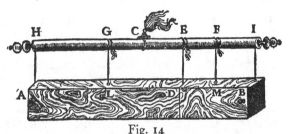

Fig. 14

But circumstances would remain the same if that part of the
prism which is now held up, at the ends, by the threads AH and
DE were supported at the middle by a single thread GL; and
likewise the other part DB would not change position if held
by a thread FM placed at its middle point. Suppose now the
threads HA, ED, and IB to be removed, leaving only the two
GL and FM, then the same equilibrium will be maintained so
long as the suspension is at C. Now let us consider that we have
here two heavy bodies AD and DB hung at the ends G and F, of
a balance beam GF in equilibrium about the point C, so that
the line CG is the distance from C to the point of suspension
of the heavy body AD, while CF is the distance at which the
other heavy body, DB, is supported. It remains now only to
show that these distances bear to each other the inverse ratio
of the weights themselves, that is, the distance GC is to the
distance CF as the prism DB is to the prism DA—a proposition
which we shall prove as follows: Since the line GE is the half of
EH, and since EF is the half of EI, the whole length GF will be
half

half of the entire line HI, and therefore equal to CI: if now we subtract the common part CF the remainder GC will be equal to the remainder FI, that is, to FE, and if to each of these we add CE we shall have GE equal to CF: hence GE:EF=FC:CG. But GE and EF bear the same ratio to each other as do their doubles HE and EI, that is, the same ratio as the prism AD to DB. Therefore, by equating ratios we have, *convertendo*, the distance GC is to the distance CF as the weight BD is to the weight DA, which is what I desired to prove.

[154]

If what precedes is clear, you will not hesitate, I think, to admit that the two prisms AD and DB are in equilibrium about the point C since one-half of the whole body AB lies on the right of the suspension C and the other half on the left; in other words, this arrangement is equivalent to two equal weights disposed at equal distances. I do not see how any one can doubt, if the two prisms AD and DB were transformed into cubes, spheres, or into any other figure whatever and if G and F were retained as points of suspension, that they would remain in equilibrium about the point C, for it is only too evident that change of figure does not produce change of weight so long as the mass [*quantità di materia*] does not vary. From this we may derive the general conclusion that any two heavy bodies are in equilibrium at distances which are inversely proportional to their weights.

This principle established, I desire, before passing to any other subject, to call your attention to the fact that these forces, resistances, moments, figures, etc., may be considered either in the abstract, dissociated from matter, or in the concrete, associated with matter. Hence the properties which belong to figures that are merely geometrical and non-material must be modified when we fill these figures with matter and therefore give them weight. Take, for example, the lever BA which, resting upon the support E, is used to lift a heavy stone D. The principle just demonstrated makes it clear that a force applied at the extremity B will just suffice to equilibrate the resistance offered by the heavy body D provided this force [*momento*] bears to the force [*momento*] at D the same ratio as the distance

distance AC bears to the distance CB; and this is true so long as
we consider only the moments of the single force at B and of the
resistance at D, treating the lever as an immaterial body devoid
of weight. But if we take into account the weight of the lever
itself—an instrument which may be made either of wood or of
iron—it is manifest that, when this weight has been added to the

[155]

force at B, the ratio will be changed and must therefore be
expressed in different terms. Hence before going further let

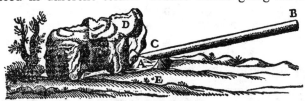

Fig. 15

us agree to distinguish between these two points of view; when
we consider an instrument in the abstract, i. e., apart from the
weight of its own material, we shall speak of "taking it in an
absolute sense" [*prendere assolutamente*]; but if we fill one of these
simple and absolute figures with matter and thus give it weight,
we shall refer to such a material figure as a "moment" or
"compound force" [*momento o forza composta*].

SAGR. I must break my resolution about not leading you off
into a digression; for I cannot concentrate my attention upon
what is to follow until a certain doubt is removed from my
mind, namely, you seem to compare the force at B with the
total weight of the stone D, a part of which—possibly the
greater part—rests upon the horizontal plane: so that . . .

SALV. I understand perfectly: you need go no further. However please observe that I have not mentioned the total weight
of the stone; I spoke only of its force [*momento*] at the point A,
the extremity of the lever BA, which force is always less than
the total weight of the stone, and varies with its shape and
elevation.

SAGR. Good: but there occurs to me another question about
which

which I am curious. For a complete understanding of this matter, I should like you to show me, if possible, how one can determine what part of the total weight is supported by the underlying plane and what part by the end A of the lever.

SALV. The explanation will not delay us long and I shall therefore have pleasure in granting your request. In the accompanying figure, let us understand that the weight having its center of gravity at A rests with the end B upon the horizontal plane and with the other end upon the lever CG. Let N be the fulcrum of a lever to which the force [*potenza*] is applied at G. Let fall the perpendiculars, AO and CF, from the center A and the end C. Then I say, the magnitude [*momento*] of the entire weight bears to the magnitude of the force [*momento della potenza*] at G a ratio compounded of the ratio between the two

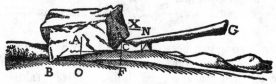

Fig. 16

distances GN and NC and the ratio between FB and BO. Lay off a distance X such that its ratio to NC is the same as that of BO to FB; then, since the total weight A is counterbalanced by the two forces at B and at C, it follows that the force at B is to that at C as the distance FO is to the distance OB. Hence,

[156]

componendo, the sum of the forces at B and C, that is, the total weight A [*momento di tutto 'l peso A*], is to the force at C as the line FB is to the line BO, that is, as NC is to X: but the force [*momento della potenza*] applied at C is to the force applied at G as the distance GN is to the distance NC; hence it follows, *ex æquali in proportione perturbata,** that the entire weight A is to the force applied at G as the distance GN is to X. But the ratio of GN to X is compounded of the ratio of GN to NC and of NC to X, that is, of FB to BO; hence the weight A bears to the

---

* For definition of *perturbata* see Todhunter's *Euclid*. Book V, Def. 20.
[*Trans.*]

equilibrating force at G a ratio compounded of that of GN to NC and of FB to BO: which was to be proved.

Let us now return to our original subject; then, if what has hitherto been said is clear, it will be easily understood that,

PROPOSITION I

A prism or solid cylinder of glass, steel, wood or other breakable material which is capable of sustaining a very heavy weight when applied longitudinally is, as previously remarked, easily broken by the transverse application of a weight which may be much smaller in proportion as the length of the cylinder exceeds its thickness.

Let us imagine a solid prism ABCD fastened into a wall at the end AB, and supporting a weight E at the other end; understand also that the wall is vertical and that the prism or cylinder is fastened at right angles to the wall. It is clear that, if the cylinder breaks, fracture will occur at the point B where the edge of the mortise acts as a fulcrum for the lever BC, to which the force is applied; the thickness of the solid BA is the other arm of the lever along which is located the resistance. This resistance opposes the separation of the part BD, lying outside the wall, from that portion lying inside. From the preceding, it follows that the magnitude [*momento*] of the force applied at C bears to the magnitude [*momento*] of the resistance, found in the thickness of the prism, i. e., in the attachment of the base BA to its contiguous parts, the same ratio which the length CB bears to half the length BA; if now we define absolute resistance to fracture

[157]

as that offered to a longitudinal pull (in which case the stretching force acts in the same direction as that through which the body is moved), then it follows that the absolute resistance of the prism BD is to the breaking load placed at the end of the lever BC in the same ratio as the length BC is to the half of AB in the case of a prism, or the semidiameter in the case of a cylinder. This is our first proposition.* Observe that in what

* The one fundamental error which is implicitly introduced into this proposition and which is carried through the entire discussion of the

has here been said the weight of the solid BD itself has been left out of consideration, or rather, the prism has been assumed to be devoid of weight. But if the weight of the prism is to be taken account of in conjunction with the weight E, we must add to the weight E one half that of the prism BD: so that if, for example, the latter weighs two p o u n d s and the weight E is t e n pounds w e must treat the weight E as if it were eleven pounds.

Fig. 17

SIMP. Why not twelve?

SALV. The weight E, my dear Simplicio, hanging at the extreme end C acts upon the lever BC with its full moment of ten pounds: so also would the solid BD if suspended at the same point exert its full moment of two pounds; but, as you know, this solid is uniformly distributed through-

Second Day consists in a failure to see that, in such a beam, there must be equilibrium between the forces of tension and compression over any cross-section. The correct point of view seems first to have been found by E. Mariotte in 1680 and by A. Parent in 1713. Fortunately this error does not vitiate the conclusions of the subsequent propositions which deal only with proportions—not actual strength—of beams. Following K. Pearson (Todhunter's *History of Elasticity*) one might say that Galileo's mistake lay in supposing the fibres of the strained beam to be inextensible. Or, confessing the anachronism, one might say that the error consisted in taking the lowest fibre of the beam as the neutral axis.

[*Trans.*]

out its entire length, BC, so that the parts which lie near the
end B are less effective than those more remote.

Accordingly if we strike a balance between the two, the
weight of the entire prism may be considered as concentrated
at its center of gravity which lies midway of the lever BC.
But a weight hung at the extremity C exerts a moment twice
as great as it would if suspended from the middle: therefore

[158]

if we consider the moments of both as located at the end C we
must add to the weight E one-half that of the prism.

SIMP. I understand perfectly; and moreover, if I mistake not,
the force of the two weights BD and E, thus disposed, would
exert the same moment as would the entire weight BD together
with twice the weight E suspended at the middle of the lever
BC.

SALV. Precisely so, and a fact worth remembering. Now
we can readily understand

PROPOSITION II

How and in what proportion a rod, or rather a prism, whose
width is greater than its thickness offers more resistance to
fracture when the
force is applied in
the direction of its
breadth than in the
direction of its
thickness.

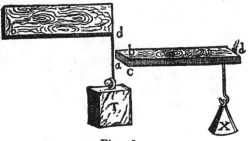

For the sake of
clearness, take a
r u l e r *ad* whose
width is *ac* a n d
whose thickness,

Fig. 18

cb, is much less than its width. The question now is why will
the ruler, if stood on edge, as in the first figure, withstand a
great weight T, while, when laid flat, as in the second figure,
it will not support the weight X which is less than T. The
answer is evident when we remember that in the one case
the

the fulcrum is at the line *bc*, and in the other case at *ca*, while the distance at which the force is applied is the same in both cases, namely, the length *bd*: but in the first case the distance of the resistance from the fulcrum—half the line *ca*—is greater than in the other case where it is only half of *bc*. Therefore the weight T is greater than X in the same ratio as half the width *ca* is greater than half the thickness *bc*, since the former acts as a lever arm for *ca*, and the latter for *cb*, against the same resistance, namely, the strength of all the fibres in the cross-section *ab*. We conclude, therefore, that any given ruler, or prism, whose width exceeds its thickness, will offer greater resistance to fracture when standing on edge than when lying flat, and this in the ratio of the width to the thickness.

PROPOSITION III

Considering now the case of a prism or cylinder growing longer in a horizontal direction, we must find out in what ratio the moment of its own weight increases in comparison with its resistance to fracture. This moment I find increases in propor-

[159]

tion to the square of the length. In order to prove this let AD be a prism or cylinder lying horizontal with its end A firmly fixed in a wall. Let the length of the prism be increased by the addition of the portion BE. It is clear that merely changing the length of the lever from AB to AC will, if we disregard its weight, increase the moment of the force [at the end] tending to produce fracture at A in the ratio of CA to BA. But, besides this, the weight of the solid portion BE, added to the weight of the solid AB increases the moment of the total weight in the ratio of the weight of the prism AE to that of the prism AB, which is the same as the ratio of the length AC to AB.

It follows, therefore, that, when the length and weight are simultaneously increased in any given proportion, the moment, which is the product of these two, is increased in a ratio which is the square of the preceding proportion. The conclusion is then that the bending moments due to the weight of prisms and cylinders which have the same thickness but different lengths, bear

bear to each other a ratio which is the square of the ratio of their lengths, or, what is the same thing, the ratio of the squares of their lengths.

We shall next show in what ratio the resistance to fracture

Fig. 19

[bending strength], in prisms and cylinders, increases with in-
[160]
crease of thickness while the length remains unchanged. Here I say that

PROPOSITION IV

In prisms and cylinders of equal length, but of unequal thicknesses, the resistance to fracture increases in the same ratio as the cube of the diameter of the thickness, i. e., of the base.

Let A and B be two cylinders of equal lengths DG, FH; let their bases be circular but unequal, having the diameters CD and EF. Then I say that the resistance to fracture offered by the cylinder

B

B is to that offered by A as the cube of the diameter FE is to the cube of the diameter DC. For, if we consider the resistance to fracture by longitudinal pull as dependent upon the bases, i. e., upon the circles EF and DC, no one can doubt that the strength [*resistenza*] of the cylinder B is greater than that of A in the same proportion in which the area of the circle EF exceeds that of CD; because it is precisely in this ratio that the number of fibres binding the parts of the solid together in the one cylinder exceeds that in the other cylinder.

But in the case of a force acting transversely it must be remembered that we are employing two levers in which the forces

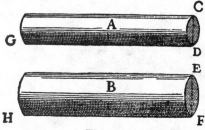

are applied at distances DG, FH, and the fulcrums are located at the points D and F; but the resistances are applied at distances which are equal to the radii of the circles DC and EF, since the fibres distributed over these entire cross-sections act as if concentrated at the centers. Remembering this and remembering also that the arms, DG and FH, through which the forces G and H act are equal, we can understand that the resistance, located at the center of the base EF, acting against the force at H, is more effective [*maggiore*] than the resistance at the center of the base CD opposing the force G, in the ratio of the radius FE to the radius DC. Accordingly the resistance to fracture offered by the cylinder B is greater than that of the cylinder A in a ratio which is compounded of that of the area of the circles EF and DC and that of their radii, i. e., of their diameters; but the areas of circles are as the squares of their diameters. Therefore the ratio of the resistances, being the product of the two preceding ratios, is the same as that of the cubes of the diameters. This is what I set out to prove. Also since the volume of a cube

[161]

varies as the third power of its edge we may say that the resistance

Fig. 20

sistance [strength] of a cylinder whose length remains constant varies as the third power of its diameter.

From the preceding we are able to conclude that

COROLLARY

The resistance [strength] of a prism or cylinder of constant length varies in the sesquialteral ratio of its volume.

This is evident because the volume of a prism or cylinder of constant altitude varies directly as the area of its base, i. e., as the square of a side or diameter of this base; but, as just demonstrated, the resistance [strength] varies as the cube of this same side or diameter. Hence the resistance varies in the sesquialteral ratio of the volume—consequently also of the weight—of the solid itself.

Simp. Before proceeding further I should like to have one of my difficulties removed. Up to this point you have not taken into consideration a certain other kind of resistance which, it appears to me, diminishes as the solid grows longer, and this is quite as true in the case of bending as in pulling; it is precisely thus that in the case of a rope we observe that a very long one is less able to support a large weight than a short one. Whence, I believe, a short rod of wood or iron will support a greater weight than if it were long, provided the force be always applied longitudinally and not transversely, and provided also that we take into account the weight of the rope itself which increases with its length.

Salv. I fear, Simplicio, if I correctly catch your meaning, that in this particular you are making the same mistake as many others; that is if you mean to say that a long rope, one of perhaps 40 cubits, cannot hold up so great a weight as a shorter length, say one or two cubits, of the same rope.

Simp. That is what I meant, and as far as I see the proposition is highly probable.

Salv. On the contrary, I consider it not merely improbable but false; and I think I can easily convince you of your error. Let AB represent the rope, fastened at the upper end A: at the lower end attach a weight C whose force is just sufficient to
break

break the rope. Now, Simplicio, point out the exact place where you think the break ought to occur.

[162]

SIMP. Let us say D.

SALV. And why at D?

SIMP. Because at this point the rope is not strong enough to support, say, 100 pounds, made up of the portion of the rope DB and the stone C.

SALV. Accordingly whenever the rope is stretched [*violentata*] with the weight of 100 pounds at D it will break there.

SIMP. I think so.

SALV. But tell me, if instead of attaching the weight at the end of the rope, B, one fastens it at a point nearer D, say, at E: or if, instead of fixing the upper end of the rope at A, one fastens it at some point F, just above D, will not the rope, at the point D, be subject to the same pull of 100 pounds?

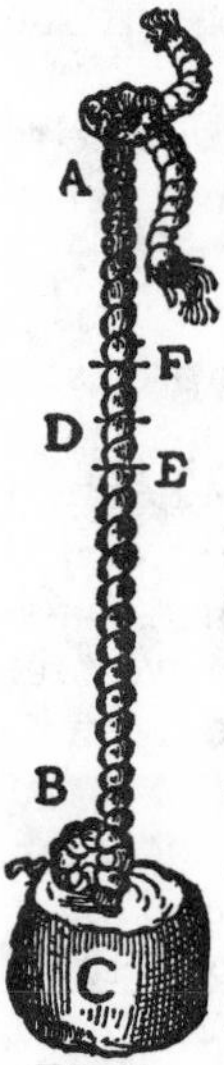

SIMP. It would, provided you include with the stone C the portion of rope EB.

SALV. Let us therefore suppose that the rope is stretched at the point D with a weight of 100 pounds, then according to your own admission it will break; but FE is only a small portion of AB; how can you therefore maintain that the long rope is weaker than the short one? Give up then this erroneous view which you share with many very intelligent people, and let us proceed.

Now having demonstrated that, in the case of [uniformly loaded] prisms and cylinders of constant thickness, the moment of force tending to produce fracture [*momento sopra le proprie resistenze*] varies

Fig. 21

as the square of the length; and having likewise shown that, when the length is constant and the thickness varies, the resistance to fracture varies as the cube of the side, or diameter, of the base, let us pass to the investigation of the case of solids which simultaneously vary in both length and thickness. Here I observe that,

PROPOSITION V

Prisms and cylinders which differ in both length and thickness offer resistances to fracture [i. e., can support at their ends loads] which are directly proportional to the cubes of the diameters of their bases and inversely proportional to their lengths.

[163]

Let ABC and DEF be two such cylinders; then the resistance [bending strength] of the cylinder AC bears to the resistance of the cylinder DF a ratio which is the product of the cube of the diameter AB divided by the cube of the diameter DE, and of the length EF divided by the A length BC.   Make EG equal to BC: let H be a third proportional to the lines AB and DE; let I be a fourth proportional, [AB/DE = H/I]:   and let I:S = EF:BC.

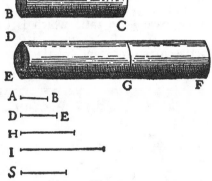

Fig. 22

Now since the resistance of the cylinder AC is to that of the cylinder DG as the cube of AB is to the cube of DE, that is, as the length AB is to the length I; and since the resistance of the cylinder DG is to that of the cylinder DF as the length FE is to EG, that is, as I is to S, it follows that the length AB is to S as the resistance of the cylinder AC is to that of the cylinder DF.   But the line AB bears to S a ratio which is the product of AB/I and I/S.   Hence the resistance [bending strength] of the cylinder AC bears to the resistance of the cylinder DF a ratio which is the product of AB/I (that is, AB³/DE³) and of I/S (that is, EF/BC): which is what I meant to prove.

This proposition having been demonstrated, let us next consider

consider the case of prisms and cylinders which are similar. Concerning these we shall show that,

PROPOSITION VI

In the case of similar cylinders and prisms, the moments [stretching forces] which result from multiplying together their weight and length [i. e., from the moments produced by their own weight and length], which latter acts as a lever-arm, bear to each other a ratio which is the sesquialteral of the ratio between the resistances of their bases.

In order to prove this let us indicate the two similar cylinders by AB and CD: then the magnitude of the force [*momento*] in the cylinder AB, opposing the resistance of its base B, bears to the magnitude [*momento*] of the force at CD, opposing the resistance of its base D, a ratio which is the sesquialteral of the ratio

[164]

between the resistance of the base B and the resistance of the

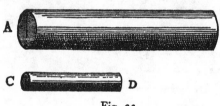

Fig. 23

base D. And since the solids AB and CD, are effective in opposing the resistances of their bases B and D, in proportion to their weights and to the mechanical advantages [*forze*] of their lever arms respectively, and since the advantage [*forza*] of the lever arm AB is equal to the advantage [*forza*] of the lever arm CD (this is true because in virtue of the similarity of the cylinders the length AB is to the radius of the base B as the length CD is to the radius of the base D), it follows that the total force [*momento*] of the cylinder AB is to the total force [*momento*] of the cylinder CD as the weight alone of the cylinder AB is to the weight alone of the cylinder CD, that is, as the volume of the cylinder AB [*l'istesso cilindro AB*] is to the volume CD [*all'istesso CD*]: but these are as the cubes of the diameters of their bases B and D; and the resistances of the bases, being

to

to each other as their areas, are to each other consequently as the squares of their diameters. Therefore the forces [*momenti*] of the cylinders are to each other in the sesquialteral ratio of the resistance of their bases.*

SIMP. This proposition strikes me as both new and surprising: at first glance it is very different from anything which I myself should have guessed: for since these figures are similar in all other respects, I should have certainly thought that the forces [*momenti*] and the resistances of these cylinders would have borne to each other the same ratio.

SAGR. This is the proof of the proposition to which I referred, at the very beginning of our discussion, as one imperfectly understood by me.

SALV. For a while, Simplicio, I used to think, as you do, that the resistances of similar solids were similar; but a certain casual observation showed me that similar solids do not exhibit a strength which is proportional to their size, the larger ones being less fitted to undergo rough usage just as tall men are more apt than small children to be injured by a fall. And, as we remarked at the outset, a large beam or column falling from a

[165]

given height will go to pieces when under the same circumstances a small scantling or small marble cylinder will not break. It was this observation which led me to the investigation of the fact which I am about to demonstrate to you: it is a very remarkable thing that, among the infinite variety of solids which are similar one to another, there are no two of which the forces [*momenti*], and the resistances of these solids are related in the same ratio.

SIMP. You remind me now of a passage in Aristotle's *Questions*

---

* The preceding paragraph beginning with Prop. VI is of more than usual interest as illustrating the confusion of terminology current in the time of Galileo. The translation given is literal except in the case of those words for which the Italian is supplied. The facts which Galileo has in mind are so evident that it is difficult to see how one can here interpret "*moment*" to mean the force "*opposing the resistance of its base,*" unless "*the force of the lever arm AB*" be taken to mean "*the mechanical advantage of the lever made up of AB and the radius of the base B*"; and similarly for "*the force of the lever arm CD.*"

[*Trans.*]

in Mechanics in which he tries to explain why it is that a wooden beam becomes weaker and can be more easily bent as it grows longer, notwithstanding the fact that the shorter beam is thinner and the longer one thicker: and, if I remember correctly, he explains it in terms of the simple lever.

SALV. Very true: but, since this solution seemed to leave room for doubt, Bishop di Guevara,* whose truly learned commentaries have greatly enriched and illuminated this work, indulges in additional clever speculations with the hope of thus overcoming all difficulties; nevertheless even he is confused as regards this particular point, namely, whether, when the length and thickness of these solid figures increase in the same ratio, their strength and resistance to fracture, as well as to bending, remain constant. After much thought upon this subject, I have reached the following result. First I shall show that,

PROPOSITION VII

Among heavy prisms and cylinders of similar figure, there is one and only one which under the stress of its own weight lies just on the limit between breaking and not breaking: so that every larger one is unable to carry the load of its own weight and breaks; while every smaller one is able to withstand some additional force tending to break it.

Let AB be a heavy prism, the longest possible that will just sustain its own weight, so that if it be lengthened the least bit it will break. Then, I say, this prism is unique among all similar prisms—infinite in number—in occupying that boundary line between breaking and not breaking; so that every larger one
[166]
will break under its own weight, and every smaller one will not break, but will be able to withstand some force in addition to its own weight.

Let the prism CE be similar to, but larger than, AB: then, I say, it will not remain intact but will break under its own weight. Lay off the portion CD, equal in length to AB. And, since, the resistance [bending strength] of CD is to that of AB as

* Bishop of Teano; b. 1561; d. 1641. [*Trans.*]

the cube of the thickness of CD is to the cube of the thickness of AB, that is, as the prism CE is to the similar prism AB, it follows that the weight of CE is the utmost load which a prism of the length CD can sustain; but the length of CE is greater; therefore the prism CE will break.

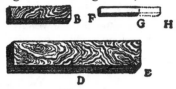

Fig. 24

Now take another prism FG which is smaller than AB. Let FH equal AB, then it can be shown in a similar manner that the resistance [bending strength] of FG is to that of AB as the prism FG is to the prism AB provided the distance AB that is FH, is equal to the distance FG; but AB is greater than FG, and therefore the moment of the prism FG applied at G is not sufficient to break the prism FG.

SAGR. The demonstration is short and clear; while the proposition which, at first glance, appeared improbable is now seen to be both true and inevitable. In order therefore to bring this prism into that limiting condition which separates breaking from not breaking, it would be necessary to change the ratio between thickness and length either by increasing the thickness or by diminishing the length. An investigation of this limiting state will, I believe, demand equal ingenuity.

SALV. Nay, even more; for the question is more difficult; this I know because I spent no small amount of time in its discovery which I now wish to share with you.

PROPOSITION VIII

Given a cylinder or prism of the greatest length consistent with its not breaking under its own weight; and having given a greater length, to find the diameter of another cylinder or prism of this greater length which shall be the only and largest one capable of withstanding its own weight.

Let BC be the largest cylinder capable of sustaining its own weight; and let DE be a length greater than AC: the problem is to find the diameter of the cylinder which, having the length

[167]

DE,

DE, shall be the largest one just able to withstand its own weight. Let I be a third proportional to the lengths DE and AC; let the diameter FD be to the diameter BA as DE is to I; draw the cylinder FE; then, among all cylinders having the same proportions, this is the largest and only one just capable of sustaining its own weight.

Let M be a third proportional to DE and I: also let O be a fourth proportional to DE, I, and M; lay off FG equal to AC. Now since the diameter FD is to the diameter AB as the length DE is to I, and since O is a fourth proportional to DE, I and M, it follows that $\overline{FD}^3:\overline{BA}^3=DE:O$. But the resistance [bending

Fig. 25

strength] of the cylinder DG is to the resistance of the cylinder BC as the cube of FD is to the cube of BA: hence the resistance of the cylinder DG is to that of cylinder BC as the length DE is to O. And since the moment of the cylinder BC is held in equilibrium by [*e equale alla*] its resistance, we shall accomplish our end (which is to prove that the moment of the cylinder FE is equal to the resistance located at FD), if we show that the moment of the cylinder FE is to the moment of the cylinder BC as the resistance DF is to the resistance BA, that is, as the cube of FD is to the cube of BA, or as the length DE is to O. The moment of the cylinder FE is to the moment of the cylinder DG as the square of DE is to the square of AC, that is, as the length DE is to I; but the moment of the cylinder DG is to the moment of the cylinder BC, as the square of DF is to the square of BA, that is, as the square of DE is to the square of I, or as the square of I is to the square of M, or, as I is to O. Therefore by equating ratios, it results that the moment of the cylinder FE is to the moment of the cylinder BC as the length DE is to O, that is, as the cube of DF is to the cube of BA, or as the resistance of the base DF is to the resistance of the base BA; which was to be proven.

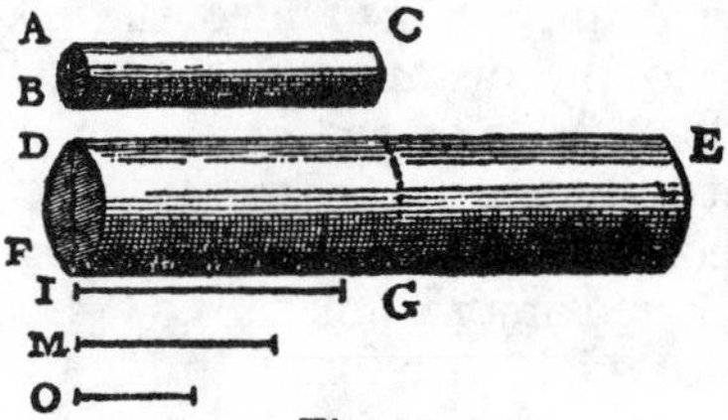

SAGR. This demonstration, Salviati, is rather long and difficult

cult to keep in mind from a single hearing. Will you not, therefore, be good enough to repeat it?

SALV. As you like; but I would suggest instead a more direct and a shorter proof: this will, however, necessitate a different figure.

[168]

SAGR. The favor will be that much greater: nevertheless I hope you will oblige me by putting into written form the argument just given so that I may study it at my leisure.

SALV. I shall gladly do so. Let A denote a cylinder of diameter DC and the largest capable of sustaining its own weight: the problem is to determine a larger cylinder which shall be at once the maximum and the unique one capable of sustaining its own weight.

Let E be such a cylinder, similar to A, having the assigned length, and having a diameter KL. Let MN be a third proportional to the two lengths DC and KL: let MN also be the diameter of another cylinder, X, having the same length as E: then, I say, X is the cylinder sought. Now since the resistance of the base DC is to the resistance of the base KL as the square of DC is to the square of KL, that is, as the square of KL is to the square of MN, or, as the cylinder E is to the cylinder X, that is, as the moment E is to the moment X; and since also the resistance [bending strength] of the base KL is to the resistance of the base MN as the cube of KL is to the cube of MN, that is, as the cube of DC is to the cube of KL, or, as the cylinder A is to the cylinder E, that is, as the moment of A is to the moment of E; hence it follows, *ex æquali in proportione perturbata,** that the moment of A is to the moment of X as the resistance of the base DC is to the resistance of the base MN; therefore moment and resistance are related to each other in prism X precisely as they are in prism A.

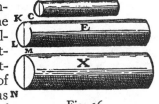

Fig. 26

* For definition of *perturbata* see Todhunter's *Euclid*, Book V, Def. 20.
[*Trans.*]

Let us now generalize the problem; then it will read as follows:

> Given a cylinder AC in which moment and resistance [bending strength] are related in any manner whatsoever; let DE be the length of another cylinder; then determine what its thickness must be in order that the relation between its moment and resistance shall be identical with that of the cylinder AC.

Using Fig. 25 in the same manner as above, we may say that, since the moment of the cylinder FE is to the moment of the portion DG as the square of ED is to the square of FG, that is, as the length DE is to I; and since the moment of the cylinder FG is to the moment of the cylinder AC as the square of FD is to the square of AB, or, as the square of ED is to the square of I, or, as the square of I is to the square of M, that is, as the length I is to O; it follows, *ex æquali*, that the moment of the

[169]

cylinder FE is to the moment of the cylinder AC as the length DE is to O, that is, as the cube of DE is to the cube of I, or, as the cube of FD is to the cube of AB, that is, as the resistance of the base FD is to the resistance of the base AB; which was to be proven.

From what has already been demonstrated, you can plainly see the impossibility of increasing the size of structures to vast dimensions either in art or in nature; likewise the impossibility of building ships, palaces, or temples of enormous size in such a way that their oars, yards, beams, iron-bolts, and, in short, all their other parts will hold together; nor can nature produce trees of extraordinary size because the branches would break down under their own weight; so also it would be impossible to build up the bony structures of men, horses, or other animals so as to hold together and perform their normal functions if these animals were to be increased enormously in height; for this increase in height can be accomplished only by employing a material which is harder and stronger than usual, or by enlarging the size of the bones, thus changing their shape until the form and appearance of the animals suggest a monstrosity. This is perhaps

perhaps what our wise Poet had in mind, when he says, in describing a huge giant:

> "Impossible it is to reckon his height
> "So beyond measure is his size." *

To illustrate briefly, I have sketched a bone whose natural length has been increased three times and whose thickness has been multiplied until, for a correspondingly large animal, it would perform the same function which the small bone performs for its small animal. From the figures here shown you can see how out of proportion the enlarged bone appears. Clearly then if one wishes to maintain in a great giant the same proportion of limb as that found in an ordinary man he must either find a harder and stronger material for making the [170] bones, or he must admit a diminution of strength in comparison with men of medium stature; for if his height be increased

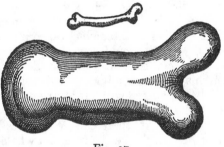

Fig. 27

inordinately he will fall and be crushed under his own weight. Whereas, if the size of a body be diminished, the strength of that body is not diminished in the same proportion; indeed the smaller the body the greater its relative strength. Thus a small dog could probably carry on his back two or three dogs of his own size; but I believe that a horse could not carry even one of his own size.

Simp. This may be so; but I am led to doubt it on account of the enormous size reached by certain fish, such as the whale which, I understand, is ten times as large as an elephant; yet they all support themselves.

Salv. Your question, Simplicio, suggests another principle,

---

* *Non si può compartir quanto sia lungo,*
 Si smisuratamente è tutto grosso.
    Ariosto's *Orlando Furioso*, XVII, 30 [*Trans.*]

one which had hitherto escaped my attention and which enables giants and other animals of vast size to support themselves and to move about as well as smaller animals do. This result may be secured either by increasing the strength of the bones and other parts intended to carry not only their weight but also the superincumbent load; or, keeping the proportions of the bony structure constant, the skeleton will hold together in the same manner or even more easily, provided one diminishes, in the proper proportion, the weight of the bony material, of the flesh, and of anything else which the skeleton has to carry. It is this second principle which is employed by nature in the structure of fish, making their bones and muscles not merely light but entirely devoid of weight.

SIMP. The trend of your argument, Salviati, is evident. Since fish live in water which on account of its density [*corpulenza*] or, as others would say, heaviness [*gravità*] diminishes the weight [*peso*] of bodies immersed in it, you mean to say that, for this reason, the bodies of fish will be devoid of weight and will be supported without injury to their bones. But this is not all; for although the remainder of the body of the fish may be without weight, there can be no question but that their bones have weight. Take the case of a whale's rib, having the dimensions of a beam; who can deny its great weight or its tendency to go to the bottom when placed in water? One would, therefore,

[171]

hardly expect these great masses to sustain themselves.

SALV. A very shrewd objection! And now, in reply, tell me whether you have ever seen fish stand motionless at will under water, neither descending to the bottom nor rising to the top, without the exertion of force by swimming?

SIMP. This is a well-known phenomenon.

SALV. The fact then that fish are able to remain motionless under water is a conclusive reason for thinking that the material of their bodies has the same specific gravity as that of water; accordingly, if in their make-up there are certain parts which are heavier than water there must be others which are lighter, for otherwise they would not produce equilibrium.

Hence

Hence, if the bones are heavier, it is necessary that the muscles or other constituents of the body should be lighter in order that their buoyancy may counterbalance the weight of the bones. In aquatic animals therefore circumstances are just reversed from what they are with land animals inasmuch as, in the latter, the bones sustain not only their own weight but also that of the flesh, while in the former it is the flesh which supports not only its own weight but also that of the bones. We must therefore cease to wonder why these enormously large animals inhabit the water rather than the land, that is to say, the air.

SIMP. I am convinced and I only wish to add that what we call land animals ought really to be called air animals, seeing that they live in the air, are surrounded by air, and breathe air.

SAGR. I have enjoyed Simplicio's discussion including both the question raised and its answer. Moreover I can easily understand that one of these giant fish, if pulled ashore, would not perhaps sustain itself for any great length of time, but would be crushed under its own mass as soon as the connections between the bones gave way.

SALV. I am inclined to your opinion; and, indeed, I almost think that the same thing would happen in the case of a very big ship which floats on the sea without going to pieces under
[172]
its load of merchandise and armament, but which on dry land and in air would probably fall apart. But let us proceed and show how:

Given a prism or cylinder, also its own weight and the maximum load which it can carry, it is then possible to find a maximum length beyond which the cylinder cannot be prolonged without breaking under its own weight.

Let AC indicate both the prism and its own weight; also let D represent the maximum load which the prism can carry at the end C without fracture; it is required to find the maximum to which the length of the said prism can be increased without breaking. Draw AH of such a length that the weight of the prism AC is to the sum of AC and twice the weight D

as

as the length CA is to AH; and let AG be a mean proportional between CA and AH; then, I say, AG is the length sought. Since the moment of the weight [*momento gravante*] D attached at the point C is equal to the moment of a weight twice as large as D placed at the middle point AC, through which the weight of

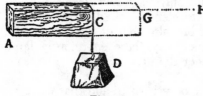

Fig. 28

the prism AC acts, it follows that the moment of the resistance of the prism AC located at A is equivalent to twice the weight D plus the weight of AC, both acting through the middle point of AC. And since we have agreed that the moment of the weights thus located, namely, twice D plus AC, bears to the moment of AC the same ratio which the length HA bears to CA and since AG is a mean proportional between these two lengths, it follows that the moment of twice D plus AC is to the moment of AC as the square of GA is to the square of CA. But the moment arising from the weight [*momento premente*] of the prism GA is to the moment of AC as the square of GA is to the square of CA; thence AG is the maximum length sought, that is, the length up to which the prism AC may be prolonged and still support itself, but beyond which it will break.

Hitherto we have considered the moments and resistances of prisms and solid cylinders fixed at one end with a weight applied at the other end; three cases were discussed, namely, that in which the applied force was the only one acting, that in which the weight of the prism itself is also taken into consideration, and that in which the weight of the prism alone is taken into consideration.   Let us now consider these same

[173]

prisms and cylinders when supported at both ends or at a single point placed somewhere between the ends.   In the first place, I remark that a cylinder carrying only its own weight and having the maximum length, beyond which it will break, will, when supported either in the middle or at both ends, have twice the length

length of one which is mortised into a wall and supported only at one end. This is very evident because, if we denote the cylinder by ABC and if we assume that one-half of it, AB, is the greatest possible length capable of supporting its own weight with one end fixed at B, then, for the same reason, if the cylinder is carried on the point G, the first half will be counterbalanced by the other half BC. So also in the case of the cylinder DEF, if its length be such that it will support only one-half this

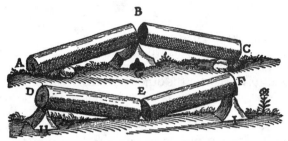

Fig. 29

length when the end D is held fixed, or the other half when the end F is fixed, then it is evident that when supports, such as H and I, are placed under the ends D and F respectively the moment of any additional force or weight placed at E will produce fracture at this point.

A more intricate and difficult problem is the following: neglect the weight of a solid such as the preceding and find whether the same force or weight which produces fracture when applied at the middle of a cylinder, supported at both ends, will also break the cylinder when applied at some other point nearer one end than the other.

Thus, for example, if one wished to break a stick by holding it with one hand at each end and applying his knee at the middle, would the same force be required to break it in the same manner if the knee were applied, not at the middle, but at some point nearer to one end?

Sagr. This problem, I believe, has been touched upon by Aristotle in his *Questions in Mechanics*.

Salv.

SALV. His inquiry however is not quite the same; for he seeks merely to discover why it is that a stick may be more easily broken by taking hold, one hand at each end of the stick, that is, far removed from the knee, than if the hands were closer together. He gives a general explanation, referring it to the lengthened lever arms which are secured by placing the hands at the ends of the stick. Our inquiry calls for something more: what we want to know is whether, when the hands are retained at the ends of the stick, the same force is required to break it wherever the knee be placed.

SAGR. At first glance this would appear to be so, because the two lever arms exert, in a certain way, the same moment, seeing that as one grows shorter the other grows correspondingly longer.

SALV. Now you see how readily one falls into error and what caution and circumspection are required to avoid it. What you have just said appears at first glance highly probable, but on closer examination it proves to be quite far from true; as will be seen from the fact that whether the knee—the fulcrum of the two levers—be placed in the middle or not makes such a difference that, if fracture is to be produced at any other point than the middle, the breaking force at the middle, even when multiplied four, ten, a hundred, or a thousand times would not suffice. To begin with we shall offer some general considerations and then pass to the determination of the ratio in which the breaking force must change in order to produce fracture at one point rather than another.

Let AB denote a wooden cylinder which is to be broken in the middle, over the supporting point C, and let DE represent an identical cylinder which is to be broken just over the supporting point F which is not in the middle. First of all it is clear that, since the distances AC and CB are equal, the forces applied at the extremities B and A must also be equal. Secondly since the distance DF is less than the distance AC the moment of any force acting at D is less than the moment of the same force at A, that is, applied at the distance CA; and the moments are less in the ratio of the length DF to AC; consequently it is

necessary

necessary to increase the force [*momento*] at D in order to overcome, or even to balance, the resistance at F; but in comparison
with the length AC the distance DF can be diminished indefinitely: in order therefore to counterbalance the resistance at
F it will be necessary to increase indefinitely the force [*forza*]
applied at D. On the other
hand, in proportion as we in-
[175]
crease the distance FE over
that of CB, we must diminish
the force at E in order to
counterbalance the resistance
at F; but the distance FE,
measured in terms of CB,
cannot be increased indefi-

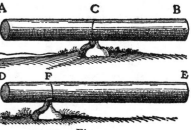

Fig. 30

nitely by sliding the fulcrum F toward the end D; indeed, it cannot even be made double the length CB. Therefore the force required at E to balance the resistance at F will always be more
than half that required at B. It is clear then that, as the fulcrum
F approaches the end D, we must of necessity indefinitely increase the sum of the forces applied at E and D in order to
balance, or overcome, the resistance at F.

SAGR. What shall we say, Simplicio? Must we not confess
that geometry is the most powerful of all instruments for
sharpening the wit and training the mind to think correctly?
Was not Plato perfectly right when he wished that his pupils
should be first of all well grounded in mathematics? As for
myself, I quite understood the property of the lever and how,
by increasing or diminishing its length, one can increase or
diminish the moment of force and of resistance; and yet, in
the solution of the present problem I was not slightly, but greatly,
deceived.

SIMP. Indeed I begin to understand that while logic is an excellent guide in discourse, it does not, as regards stimulation to
discovery, compare with the power of sharp distinction which
belongs to geometry.

SAGR. Logic, it appears to me, teaches us how to test the
conclusiveness

conclusiveness of any argument or demonstration already discovered and completed; but I do not believe that it teaches
us to discover correct arguments and demonstrations. But it
would be better if Salviati were to show us in just what proportion the forces must be increased in order to produce fracture
as the fulcrum is moved from one point to another along one
and the same wooden rod.

[176]

SALV. The ratio which you desire is determined as follows:
If upon a cylinder one marks two points at which fracture is to be produced, then the resistances at these two
points will bear to each other the inverse ratio of the
rectangles formed by the distances from the respective
points to the ends of the cylinder.

Let A and B denote the least forces which will bring about
fracture of the cylinder at C; likewise E and F the smallest
forces which will break it at D. Then, I say, that the sum of the
forces A and B is to the sum of the forces E and F as the area
of the rectangle AD.DB is to the area of the rectangle AC.CB.
Because the sum of the forces A and B bears to the sum of the
forces E and F a ratio which is the product of the three following
ratios, namely, $(A+B)/B$, B/F, and $F/(F+E)$; but the length
BA is to the length CA as the sum of the forces A and B is to the

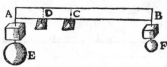

Fig. 31

force B; and, as the length DB
is to the length CB, so is the
force B to the force F; also as
the length AD is to AB, so is the
force F to the sum of the forces F
and E.

Hence it follows that the sum of the forces A and B bears
to the sum of the forces E and F a ratio which is the product
of the three following ratios, namely, BA/CA, BD/BC, and
AD/AB. But DA/CA is the product of DA/BA and BA/CA.
Therefore the sum of the forces A and B bears to the sum of the
forces E and F a ratio which is the product of DA:CA and
DB:CB. But the rectangle AD.DB bears to the rectangle
AC.CB a ratio which is the product of DA/CA and DB/CB.
Accordingly

Accordingly the sum of the forces A and B is to the sum of the forces E and F as the rectangle AD.DB is to the rectangle AC.CB, that is, the resistance to fracture at C is to the resistance to fracture at D as the rectangle AD.DB is to the rectangle AC.CB. Q. E. D.

[177]

Another rather interesting problem may be solved as a consequence of this theorem, namely,

Given the maximum weight which a cylinder or prism can support at its middle-point where the resistance is a minimum, and given also a larger weight, find that point in the cylinder for which this larger weight is the maximum load that can be supported.

Let that one of the given weights which is larger than the maximum weight supported at the middle of the cylinder AB bear to this maximum weight the same ratio which the length E bears to the length F. The problem is to find that point in the cylinder at which this larger weight becomes the maximum that can be supported. Let G be a mean proportional between the lengths E and F. Draw AD and S so that they bear to each other the same ratio as E to G; accordingly S will be less than AD.

Let AD be the diameter of a semicircle AHD, in which take AH equal to S; join the points H and D and lay off DR equal to HD. Then, I say, R is the point sought, namely, the point at which the given weight, greater than the maximum supported at the middle of the cylinder D, would become the maximum load.

On AB as diameter draw the semicircle ANB: erect the perpendicular RN and join the points N and D. Now since the sum of the squares on NR and RD is equal to the square of ND, that is, to the square of AD, or to the sum of the squares of AH and HD; and, since the square of HD is equal to the square of DR, it follows that the square of NR, that is, the rectangle AR.RB, is equal to the square of AH, also therefore to the square of S; but the square of S is to the square of AD as the length F is to the length E, that is, as the maximum weight supported

supported at D is to the larger of the two given weights. Hence the latter will be the maximum load which can be carried at the point R; which is the solution sought.

SAGR. Now I understand thoroughly; and I am thinking that, since the prism AB grows constantly stronger and more resistant

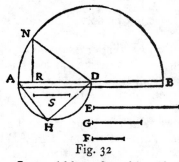

Fig. 32

to the pressure of its load at points which are more and more removed from the middle, we could in the case of large heavy beams cut away a considerable portion near the ends which would notably lessen the weight, and which, in the beam work of large rooms, would prove to be of great utility and convenience. [178]

It would be a fine thing if one could discover the proper shape to give a solid in order to make it equally resistant at every point, in which case a load placed at the middle would not produce fracture more easily than if placed at any other point.*

SALV. I was just on the point of mentioning an interesting and remarkable fact connected with this very question. My meaning will be clearer if I draw a figure. Let DB represent a prism; then, as we have already shown, its resistance to fracture [bending strength] at the end AD, owing to a load placed at the end B, will be less than the resistance at CI in the ratio of the length CB to AB. Now imagine this same prism to be cut through diagonally along the line FB so that the opposite faces will be triangular; the side facing us will be FAB. Such a solid

---

* The reader will notice that two different problems are here involved. That which is suggested in the last remark of Sagredo is the following:

   To find a beam whose maximum stress has the same value when a constant load moves from one end of the beam to the other.

   The second problem—the one which Salviati proceeds to solve—is the following:

   To find a beam in all cross-sections of which the maximum stress is the same for a constant load in a fixed position. [*Trans.*]

will have properties different from those of the prism; for, if the load remain at B, the resistance against fracture [bending strength] at C will be less than that at A in the ratio of the length CB to the length AB. This is easily proved: for if CNO represents a cross-section parallel to AFD, then the length FA bears to the length CN, in the triangle FAB, the same ratio which the length AB bears to the length CB. Therefore, if we imagine A and C to be the points at which the fulcrum is placed, the lever arms in the two cases BA, AF and BC, CN will be proportional [*simili*].

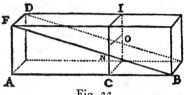

Fig. 33

Hence the moment of any force applied at B and acting through the arm BA, against a resistance placed at a distance AF will be equal to that of the same force at B acting through the arm BC against the same resistance located at a distance CN. But now, if the force still be applied at B, the resistance to be overcome when the fulcrum is at C, acting through the arm CN, is less than the resistance with the fulcrum at A in the same proportion as the rectangular cross-section CO is less than the rectangular cross-section AD, that is, as the length CN is less than AF, or CB than BA.

Consequently the resistance to fracture at C, offered by the portion OBC, is less than the resistance to fracture at A, offered by the entire block DAB, in the same proportion as the length CB is smaller than the length AB.

By this diagonal saw-cut we have now removed from the beam, or prism DB, a portion, i. e., a half, and have left the wedge, or triangular prism, FBA. We thus have two solids
[179]
possessing opposite properties; one body grows stronger as it is shortened while the other grows weaker. This being so it would seem not merely reasonable, but inevitable, that there exists a line of section such that, when the superfluous material has been removed, there will remain a solid of such figure that it will offer the same resistance [strength] at all points.

Simp.

Simp. Evidently one must, in passing from greater to less, encounter equality.

Sagr. But now the question is what path the saw should follow in making the cut.

Simp. It seems to me that this ought not to be a difficult task: for if by sawing the prism along the diagonal line and removing half of the material, the remainder acquires a property just the opposite to that of the entire prism, so that at every point where the latter gains strength the former becomes weaker, then it seems to me that by taking a middle path, i. e., by removing half the former half, or one-quarter of the whole, the strength of the remaining figure will be constant at all those points where, in the two previous figures, the gain in one was equal to the loss in the other.

Salv. You have missed the mark, Simplicio. For, as I shall presently show you, the amount which you can remove from the prism without weakening it is not a quarter but a third. It now remains, as suggested by Sagredo, to discover the path along which the saw must travel: this, as I shall prove, must be a parabola. But it is first necessary to demonstrate the following lemma:

If the fulcrums are so placed under two levers or balances that the arms through which the forces act are to each other in the same ratio as the squares of the arms through which the resistances act, and if these resistances are to each other in the same ratio as the arms through which they act, then the forces will be equal.

Let AB and CD represent two levers whose lengths are

Fig. 34

divided by their fulcrums in such a way as to make the distance EB bear to the distance FD a ratio which is equal to the square of the ratio between the distances EA and FC. Let the resistances located at A and C

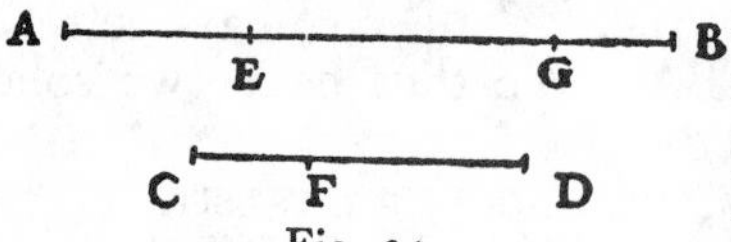

[180]

be to each other as EA is to FC. Then, I say, the forces which must be applied at B and D in order to hold in equilibrium the resistances

resistances at A and C are equal. Let EG be a mean proportional between EB and FD. Then we shall have BE:EG= EG:FD=AE:CF. But this last ratio is precisely that which we have assumed to exist between the resistances at A and C. And since EG:FD=AE:CF, it follows, *permutando*, that EG: AE=FD:CF. Seeing that the distances DC and GA are divided in the same ratio by the points F and E, it follows that the same force which, when applied at D, will equilibrate the resistance at C, would if applied at G equilibrate at A a resistance equal to that found at C.

But one datum of the problem is that the resistance at A is to the resistance at C as the distance AE is to the distance CF, or as BE is to EG. Therefore the force applied at G, or rather at D, will, when applied at B, just balance the resistance located at A.

Q. E. D.

This being clear draw the parabola FNB in the face FB of the prism DB. Let the prism be sawed along this parabola whose vertex is at B. The portion of the solid which remains will be included between the base AD, the rectangular plane AG, the straight line BG and the surface DGBF, whose curvature is identical with that of the parabola FNB. This solid will have, I say, the same strength at every point. Let the solid be cut by a plane CO parallel to the plane AD. Imagine the points A and C to be the fulcrums of two levers of which one will have the arms BA and AF; the other BC and CN. Then since in

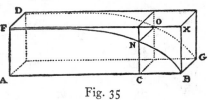

Fig. 35

the parabola FBA, we have BA:BC=$\overline{AF}^2$: $\overline{CN}^2$, it is clear that the arm BA of one lever is to the arm BC of the other lever as the square of the arm AF is to the square of the other arm CN. Since the resistance to be balanced by the lever BA is to the resistance to be balanced by the lever BC in the same ratio as the rectangle DA is to the rectangle OC, that is as the length AF is to the length CN, which two lengths are the other arms of the levers, it follows, by the lemma just demonstrated, that the

the same force which, when applied at BG will equilibrate the resistance at DA, will also balance the resistance at CO.   The

[181]

same is true for any other section.   Therefore this parabolic solid is equally strong throughout.

It can now be shown that, if the prism be sawed along the line of the parabola FNB, one-third part of it will be removed; because the rectangle FB and the surface FNBA bounded by the parabola are the bases of two solids included between two parallel planes, i. e., between the rectangles FB and DG; consequently the volumes of these two solids bear to each other the same ratio as their bases.   But the area of the rectangle is one and a half times as large as the area FNBA under the parabola; hence by cutting the prism along the parabola we remove one-third of the volume.   It is thus seen how one can diminish the weight of a beam by as much as thirty-three per cent without diminishing its strength; a fact of no small utility in the construction of large vessels, and especially in supporting the decks, since in such structures lightness is of prime importance.

SAGR. The advantages derived from this fact are so numerous that it would be both wearisome and impossible to mention them all; but leaving this matter to one side, I should like to learn just how it happens that diminution of weight is possible in the ratio above stated.   I can readily understand that, when a section is made along the diagonal, one-half the weight is removed; but, as for the parabolic section removing one-third of the prism, this I can only accept on the word of Salviati who is always reliable; however I prefer first-hand knowledge to the word of another.

SALV. You would like then a demonstration of the fact that the excess of the volume of a prism over the volume of what we have called the parabolic solid is one-third of the entire prism.   This I have already given you on a previous occasion; however I shall now try to recall the demonstration in which I remember having used a certain lemma from Archimedes' book *On Spirals,** namely, Given any number of lines, differing in

* For demonstration of the theorem here cited, see "*Works of Arch-*

length one from another by a common difference which is equal to the shortest of these lines; and given also an equal number of lines each of which has the same length as the longest of the first-mentioned series; then the sum of the squares of the lines of this second group will be less than three times the sum of the squares of the lines in the first group. But the sum of the squares of the second group will be greater than three times the sum of the squares of all excepting the longest of the first group.

[182]

Assuming this, inscribe in the rectangle ACBP the parabola AB. We have now to prove that the mixed triangle BAP whose sides are BP and PA, and whose base is the parabola BA, is a third part of the entire rectangle CP. If this is not true it will be either greater or less than a third. Suppose it to be less by an area which is represented by X. By drawing lines parallel to the sides BP and CA, we can divide the rectangle CP into equal parts; and if the process be continued we shall finally reach a division into parts so small that each of them will be smaller than the area X; let the rec-

tangle OB represent one of these parts and, through the points where the other parallels cut the parabola, draw lines parallel to AP. Let us now describe about our "mixed triangle" a figure made up of rectangles such as BO, IN, HM, FL, EK, and GA; this figure will also be less than

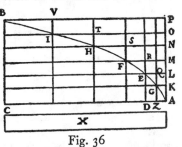

Fig. 36

a third part of the rectangle CP because the excess of this figure above the area of the "mixed triangle" is much smaller than the rectangle BO which we have already made smaller than X.

SAGR. More slowly, please; for I do not see how the excess of this figure described about the "mixed triangle" is much smaller than the rectangle BO.

SALV. Does not the rectangle BO have an area which is equal to the sum of the areas of all the little rectangles through which

imedes" translated by T. L. Heath (Camb. Univ. Press 1897) p. 107 and p. 162. [*Trans.*]

the parabola passes? I mean the rectangles BI, IH, HF, FE, EG, and GA of which only a part lies outside the "mixed triangle." Have we not taken the rectangle BO smaller than the area X? Therefore if, as our opponent might say, the triangle plus X is equal to a third part of this rectangle CP, the circumscribed figure, which adds to the triangle an area less than X, will still remain smaller than a third part of the rectangle, CP. But this cannot be, because this circumscribed figure is larger than a third of the area. Hence it is not true that our "mixed triangle" is less than a third of the rectangle.

[183]

SAGR. You have cleared up my difficulty; but it still remains to be shown that the circumscribed figure is larger than a third part of the rectangle CP, a task which will not, I believe, prove so easy.

SALV. There is nothing very difficult about it. Since in the parabola $\overline{DE}^2:\overline{ZG}^2=DA:AZ=$ rectangle KE: rectangle AG, seeing that the altitudes of these two rectangles, AK and KL, are equal, it follows that $\overline{ED}^2:\overline{ZG}^2=\overline{LA}^2:\overline{AK}^2=$ rectangle KE: rectangle KZ. In precisely the same manner it may be shown that the other rectangles LF, MH, NI, OB, stand to one another in the same ratio as the squares of the lines MA, NA, OA, PA.

Let us now consider the circumscribed figure, composed of areas which bear to each other the same ratio as the squares of a series of lines whose common difference in length is equal to the shortest one in the series; note also that the rectangle CP is made up of an equal number of areas each equal to the largest and each equal to the rectangle OB. Consequently, according to the lemma of Archimedes, the circumscribed figure is larger than a third part of the rectangle CP; but it was also smaller, which is impossible. Hence the "mixed triangle" is not less than a third part of the rectangle CP.

Likewise, I say, it cannot be greater. For, let us suppose that it is greater than a third part of the rectangle CP and let the area X represent the excess of the triangle over the third part of the rectangle CP; subdivide the rectangle into equal rectangles and continue the process until one of these subdivisions is smaller than

than the area X. Let BO represent such a rectangle smaller than X. Using the above figure, we have in the "mixed triangle" an inscribed figure, made up of the rectangles VO, TN, SM, RL, and QK, which will not be less than a third part of the large rectangle CP.

For the "mixed triangle" exceeds the inscribed figure by a quantity less than that by which it exceeds the third part of the rectangle CP; to see that this is true we have only to remember that the excess of the triangle over the third part of the rectangle CP is equal to the area X, which is less than the rectangle BO, which in turn is much less than the excess of the triangle over the inscribed figure. For the rectangle BO is

[184]

made up of the small rectangles AG, GE, EF, FH, HI, and IB; and the excess of the triangle over the inscribed figure is less than half the sum of these little rectangles. Thus since the triangle exceeds the third part of the rectangle CP by an amount X, which is more than that by which it exceeds the inscribed figure, the latter will also exceed the third part of the rectangle, CP. But, by the lemma which we have assumed, it is smaller. For the rectangle CP, being the sum of the largest rectangles, bears to the component rectangles of the inscribed figure the same ratio which the sum of all the squares of the lines equal to the longest bears to the squares of the lines which have a common difference, after the square of the longest has been subtracted.

Therefore, as in the case of squares, the sum total of the largest rectangles, i. e., the rectangle CP, is greater than three times the sum total of those having a common difference minus the largest; but these last make up the inscribed figure. Hence the "mixed triangle" is neither greater nor less than the third part of rectangle CP; it is therefore equal to it.

SAGR. A fine, clever demonstration; and all the more so because it gives us the quadrature of the parabola, proving it to be four-thirds of the inscribed * triangle, a fact which Archimedes demonstrates by means of two different, but admirable, series of

---

* Distinguish carefully between this triangle and the "mixed triangle" above mentioned. [*Trans.*]

many propositions. This same theorem has also been recently established by Luca Valerio,* the Archimedes of our age; his demonstration is to be found in his book dealing with the centers of gravity of solids.

SALV. A book which, indeed, is not to be placed second to any produced by the most eminent geometers either of the present or of the past; a book which, as soon as it fell into the hands of our Academician, led him to abandon his own researches along these lines; for he saw how happily everything had been treated and demonstrated by Valerio.

[185]

SAGR. When I was informed of this event by the Academician himself, I begged of him to show the demonstrations which he had discovered before seeing Valerio's book; but in this I did not succeed.

SALV. I have a copy of them and will show them to you; for you will enjoy the diversity of method employed by these two authors in reaching and proving the same conclusions; you will also find that some of these conclusions are explained in different ways, although both are in fact equally correct.

SAGR. I shall be much pleased to see them and will consider it a great favor if you will bring them to our regular meeting. But in the meantime, considering the strength of a solid formed from a prism by means of a parabolic section, would it not, in view of the fact that this result promises to be both interesting and useful in many mechanical operations, be a fine thing if you were to give some quick and easy rule by which a mechanician might draw a parabola upon a plane surface?

SALV. There are many ways of tracing these curves; I will mention merely the two which are the quickest of all. One of these is really remarkable; because by it I can trace thirty or forty parabolic curves with no less neatness and precision, and in a shorter time than another man can, by the aid of a compass, neatly draw four or six circles of different sizes upon paper. I take a perfectly round brass ball about the size of a walnut and project it along the surface of a metallic mirror held

* An eminent Italian mathematician, contemporary with Galileo.
[Trans.]

in a nearly upright position, so that the ball in its motion will press slightly upon the mirror and trace out a fine sharp parabolic line; this parabola will grow longer and narrower as the angle of elevation increases. The above experiment furnishes clear and tangible evidence that the path of a projectile is a parabola; a fact first observed by our friend and demonstrated by him in his book on motion which we shall take up at our next meeting. In the execution of this method, it is advisable to slightly heat and moisten the ball by rolling in the hand in order that its trace upon the mirror may be more distinct.

[186]

The other method of drawing the desired curve upon the face of the prism is the following: Drive two nails into a wall at a convenient height and at the same level; make the distance between these nails twice the width of the rectangle upon which it is desired to trace the semiparabola. Over these two nails hang a light chain of such a length that the depth of its sag is equal to the length of the prism. This chain will assume the form of a parabola,* so that if this form be marked by points on the wall we shall have described a complete parabola which can be divided into two equal parts by drawing a vertical line through a point midway between the two nails. The transfer of this curve to the two opposing faces of the prism is a matter of no difficulty; any ordinary mechanic will know how to do it.

By use of the geometrical lines drawn upon our friend's compass,† one may easily lay off those points which will locate this same curve upon the same face of the prism.

Hitherto we have demonstrated numerous conclusions pertaining to the resistance which solids offer to fracture. As a starting point for this science, we assumed that the resistance offered by the solid to a straight-away pull was known; from this base one might proceed to the discovery of many other results and their demonstrations; of these results the number to

---

* It is now well known that this curve is not a parabola but a catenary the equation of which was first given, 49 years after Galileo's death, by James Bernoulli. [*Trans.*]

† The geometrical and military compass of Galileo, described in Nat. Ed. Vol. 2. [*Trans.*]

be found in nature is infinite. But, in order to bring our daily conference to an end, I wish to discuss the strength of hollow solids, which are employed in art—and still oftener in nature—in a thousand operations for the purpose of greatly increasing strength without adding to weight; examples of these are seen in the bones of birds and in many kinds of reeds which are light and highly resistant both to bending and breaking. For if a stem of straw which carries a head of wheat heavier than the entire stalk were made up of the same amount of material in

[187]

solid form it would offer less resistance to bending and breaking. This is an experience which has been verified and confirmed in practice where it is found that a hollow lance or a tube of wood or metal is much stronger than would be a solid one of the same length and weight, one which would necessarily be thinner; men have discovered, therefore, that in order to make lances strong as well as light they must make them hollow. We shall now show that:

In the case of two cylinders, one hollow the other solid but having equal volumes and equal lengths, their resistances [bending strengths] are to each other in the ratio of their diameters.

Let AE denote a hollow cylinder and IN a solid one of the

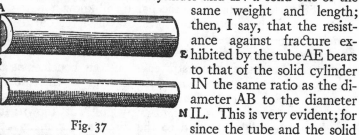

Fig. 37

same weight and length; then, I say, that the resistance against fracture exhibited by the tube AE bears to that of the solid cylinder IN the same ratio as the diameter AB to the diameter IL. This is very evident; for since the tube and the solid cylinder IN have the same volume and length, the area of the circular base IL will be equal to that of the annulus AB which is the base of the tube AE. (By annulus is here meant the area which lies between two concentric circles of different radii.) Hence their resistances to a straight-away pull are equal; but in producing

ing

ing fracture by a transverse pull we employ, in the case of the cylinder IN, the length LN as one lever arm, the point L as a fulcrum, and the diameter LI, or its half, as the opposing lever arm: while in the case of the tube, the length BE which plays the part of the first lever arm is equal to LN, the opposing lever arm beyond the fulcrum, B, is the diameter AB, or its half. Manifestly then the resistance [bending strength] of the tube exceeds that of the solid cylinder in the proportion in which the diameter AB exceeds the diameter IL· which is the desired result.

[188]

Thus the strength of a hollow tube exceeds that of a solid cylinder in the ratio of their diameters whenever the two are made of the same material and have the same weight and length.

It may be well next to investigate the general case of tubes and solid cylinders of constant length, but with the weight and the hollow portion variable. First we shall show that:

Given a hollow tube, a solid cylinder may be determined which will be equal [*eguale*] to it.

The method is very simple. Let AB denote the external and CD the internal diameter of the tube. In the larger circle lay off the line AE equal in length to the di-
ameter CD; join the points E and B. Now since the angle at E inscribed in a semicircle, AEB, is a right-angle, the area of the circle whose diameter is AB is equal to the sum of the areas of the two circles whose respective diameters are AE and EB. But AE is the diameter of the hollow portion of the tube. Therefore the area of the circle whose diameter is EB is the same as the area of the annulus ACBD. Hence a solid cylinder of circular base having a diameter EB will have the same volume as the walls of the tube of equal length.

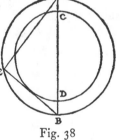

Fig. 38

By use of this theorem, it is easy:

To find the ratio between the resistance [bending strength] of any tube and that of any cylinder of equal length.

Let

Let ABE denote a tube and RSM a cylinder of equal length: it is required to find the ratio between their resistances. Using the preceding proposition, determine a cylinder ILN which shall

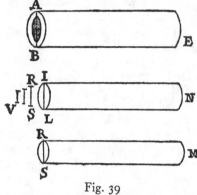

have the same volume and length as the tube. Draw a line V of such a length that it will be related to IL and RS (diameters of the bases of the cylinders IN and RM), as follows: V:RS=RS:IL. Then, I say, the resistance of the tube AE is to that of the cylinder RM as the length of the line AB is to the length

[189]

V. For, since the tube AE is equal both in volume and length, to the cylinder IN, the resistance of the tube will bear to the resistance of the cylinder the same ratio as the line AB to IL; but the resistance of the cylinder IN is to that of the cylinder RM as the cube of IL is to the cube of RS, that is, as the length IL is to length V: therefore, *ex æquali*, the resistance [bending strength] of the tube AE bears to the resistance of the cylinder RM the same ratio as the length AB to V.   Q. E. D.

Fig. 39

END OF SECOND DAY.

THIRD DAY

[190]

CHANGE OF POSITION. [*De Motu Locali*]

MY purpose is to set forth a very new science dealing with a very ancient subject. There is, in nature, perhaps nothing older than motion, concerning which the books written by philosophers are neither few nor small; nevertheless I have discovered by experiment some properties of it which are worth knowing and which have not hitherto been either observed or demonstrated. Some superficial observations have been made, as, for instance, that the free motion [*naturalem motum*] of a heavy falling body is continuously accelerated;* but to just what extent this acceleration occurs has not yet been announced; for so far as I know, no one has yet pointed out that the distances traversed, during equal intervals of time, by a body falling from rest, stand to one another in the same ratio as the odd numbers beginning with unity.†

It has been observed that missiles and projectiles describe a curved path of some sort; however no one has pointed out the fact that this path is a parabola. But this and other facts, not few in number or less worth knowing, I have succeeded in proving; and what I consider more important, there have been opened up to this vast and most excellent science, of which my

* "Natural motion" of the author has here been translated into "free motion"—since this is the term used to-day to distinguish the "natural" from the "violent" motions of the Renaissance. [*Trans.*]

† A theorem demonstrated on p. 175 below. [*Trans.*]

work is merely the beginning, ways and means by which other minds more acute than mine will explore its remote corners.

This discussion is divided into three parts; the first part deals with motion which is steady or uniform; the second treats of motion as we find it accelerated in nature; the third deals with the so-called violent motions and with projectiles.

[191]
UNIFORM MOTION

In dealing with steady or uniform motion, we need a single definition which I give as follows:

DEFINITION

By steady or uniform motion, I mean one in which the distances traversed by the moving particle during any equal intervals of time, are themselves equal.

CAUTION

We must add to the old definition (which defined steady motion simply as one in which equal distances are traversed in equal times) the word "any," meaning by this, all equal intervals of time; for it may happen that the moving body will traverse equal distances during some equal intervals of time and yet the distances traversed during some small portion of these time-intervals may not be equal, even though the time-intervals be equal.

From the above definition, four axioms follow, namely:

AXIOM I

In the case of one and the same uniform motion, the distance traversed during a longer interval of time is greater than the distance traversed during a shorter interval of time.

AXIOM II

In the case of one and the same uniform motion, the time required to traverse a greater distance is longer than the time required for a less distance.

Axiom III

In one and the same interval of time, the distance traversed at a greater speed is larger than the distance traversed at a less speed.

[192]

Axiom IV

The speed required to traverse a longer distance is greater than that required to traverse a shorter distance during the same time-interval.

Theorem I, Proposition I

If a moving particle, carried uniformly at a constant speed, traverses two distances the time-intervals required are to each other in the ratio of these distances.

Let a particle move uniformly with constant speed through two distances AB, BC, and let the time required to traverse AB be represented by DE; the time required to traverse BC, by EF;

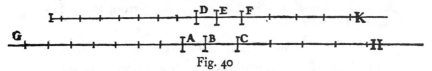

Fig. 40

then I say that the distance AB is to the distance BC as the time DE is to the time EF.

Let the distances and times be extended on both sides towards G, H and I, K; let AG be divided into any number whatever of spaces each equal to AB, and in like manner lay off in DI exactly the same number of time-intervals each equal to DE. Again lay off in CH any number whatever of distances each equal to BC; and in FK exactly the same number of time-intervals each equal to EF; then will the distance BG and the time EI be equal and arbitrary multiples of the distance BA and the time ED; and likewise the distance HB and the time KE are equal and arbitrary multiples of the distance CB and the time FE.

And since DE is the time required to traverse AB, the whole time

time EI will be required for the whole distance BG, and when the motion is uniform there will be in EI as many time-intervals each equal to DE as there are distances in BG each equal to BA; and likewise it follows that KE represents the time required to traverse HB.

Since, however, the motion is uniform, it follows that if the distance GB is equal to the distance BH, then must also the time IE be equal to the time EK; and if GB is greater than BH, then also IE will be greater than EK; and if less, less.* There

[193]

are then four quantities, the first AB, the second BC, the third DE, and the fourth EF; the time IE and the distance GB are arbitrary multiples of the first and the third, namely of the distance AB and the time DE.

But it has been proved that *both* of these latter quantities are either equal to, greater than, or less than the time EK and the space BH, which are arbitrary multiples of the second and the fourth. Therefore the first is to the second, namely the distance AB is to the distance BC, as the third is to the fourth, namely the time DE is to the time EF.   Q. E. D.

THEOREM II, PROPOSITION II

If a moving particle traverses two distances in equal intervals of time, these distances will bear to each other the same ratio as the speeds. And conversely if the distances are as the speeds then the times are equal.

Referring to Fig. 40, let AB and BC represent the two distances traversed in equal time-intervals, the distance AB for instance with the velocity DE, and the distance BC with the velocity EF. Then, I say, the distance AB is to the distance BC as the velocity DE is to the velocity EF. For if equal multiples of both distances and speeds be taken, as above, namely, GB and IE of AB and DE respectively, and in like manner HB and KE of BC and EF, then one may infer, in the same manner as above, that the multiples GB and IE are either less than, equal

* The method here employed by Galileo is that of Euclid as set forth in the famous 5th Definition of the Fifth Book of his *Elements*, for which see *art. Geometry* Ency. Brit. 11th Ed. p. 683. [*Trans.*]

to, or greater than equal multiples of BH and EK. Hence the theorem is established.

THEOREM III, PROPOSITION III

In the case of unequal speeds, the time-intervals required to traverse a given space are to each other inversely as the speeds.

Let the larger of the two unequal speeds be indicated by A; the smaller, by B; and let the motion corresponding to both traverse the given space CD. Then I say the time required to traverse the distance CD at speed A is to the time required to traverse the same distance at speed B, as the speed B is to the speed A. For let CD be to CE as A is to B; then, from the preceding, it follows that the time required to complete the distance CD at speed A is the same as the time necessary to complete CE at speed B; but the time needed to traverse the distance CE at speed B is to the time required to traverse the distance CD at the same speed as CE is to CD; therefore the time in which CD is covered at speed A is to the time in which CD is covered at speed B as CE is to CD, that is, as speed B is to speed A.   Q. E. D.

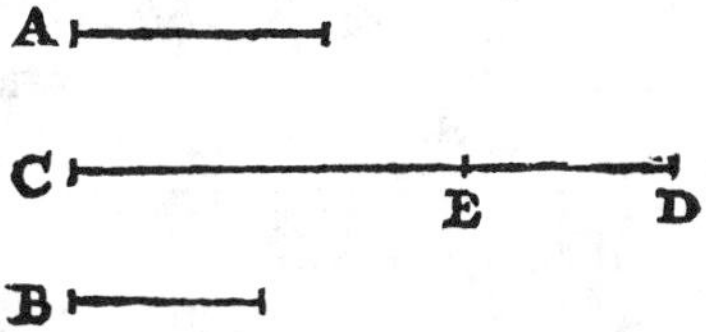

Fig. 41

[194]

THEOREM IV, PROPOSITION IV

If two particles are carried with uniform motion, but each with a different speed, the distances covered by them during unequal intervals of time bear to each other the compound ratio of the speeds and time intervals.

Let the two particles which are carried with uniform motion be E and F and let the ratio of the speed of the body E be to that of the body F as A is to B; but let the ratio of the time consumed by the motion of E be to the time consumed by the motion of F as C is to D. Then, I say, that the distance covered by E, with speed A in time C, bears to the space traversed by F with speed

B

B in time D a ratio which is the product of the ratio of the speed
A to the speed B by the ratio of the time C to the time D. For
if G is the distance traversed by E at speed A during the time-
interval C, and if G is to I as
the speed A is to the speed B;
and if also the time-interval
C is to the time-interval D
as I is to L, then it follows
that I is the distance trav-

Fig. 42

ersed by F in the same time that G is traversed by E since G
is to I in the same ratio as the speed A to the speed B. And
since I is to L in the same ratio as the time-intervals C and D,
if I is the distance traversed by F during the interval C, then
L will be the distance traversed by F during the interval D at the
speed B.

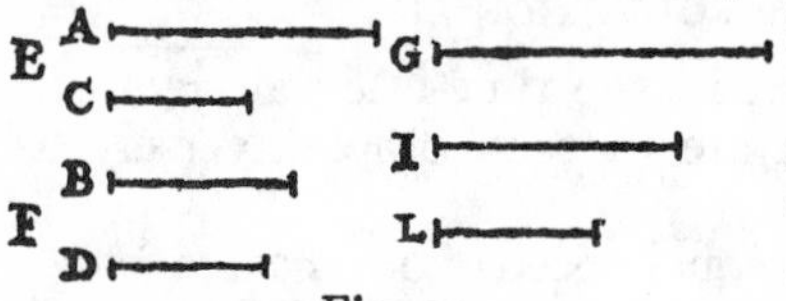

But the ratio of G to L is the product of the ratios G to I
and I to L, that is, of the ratios of the speed A to the speed B
and of the time-interval C to the time-interval D.   Q. E. D.

[195]
THEOREM V, PROPOSITION V

If two particles are moved at a uniform rate, but with unequal speeds, through unequal distances, then the ratio of
the time-intervals occupied will be the product of the ratio
of the distances by the inverse ratio of the speeds.

Let the two moving particles be denoted by A and B, and let
the speed of A be
to the speed of B in
the ratio of V to T;
in like manner let
the distances trav-
ersed be in the ratio

Fig. 43

of S to R; then I say that the ratio of the time-interval during
which the motion of A occurs to the time-interval occupied by
the motion of B is the product of the ratio of the speed T to the
speed V by the ratio of the distance S to the distance R.

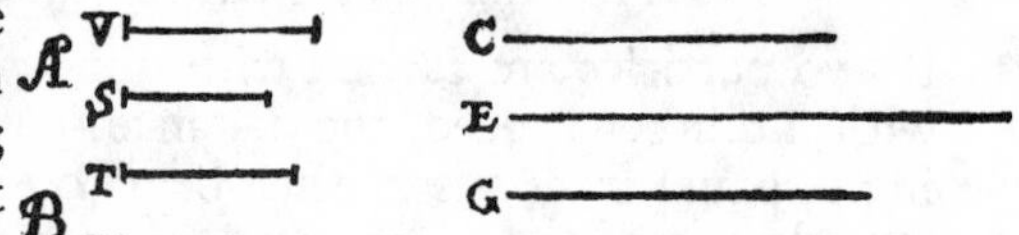

Let C be the time-interval occupied by the motion of A, and
let

let the time-interval C bear to a time-interval E the same ratio as the speed T to the speed V.

And since C is the time-interval during which A, with speed V, traverses the distance S and since T, the speed of B, is to the speed V, as the time-interval C is to the time-interval E, then E will be the time required by the particle B to traverse the distance S. If now we let the time-interval E be to the time-interval G as the distance S is to the distance R, then it follows that G is the time required by B to traverse the space R. Since the ratio of C to G is the product of the ratios C to E and E to G (while also the ratio of C to E is the inverse ratio of the speeds of A and B respectively, i. e., the ratio of T to V); and since the ratio of E to G is the same as that of the distances S and R respectively, the proposition is proved.

[196]
THEOREM VI, PROPOSITION VI

If two particles are carried at a uniform rate, the ratio of their speeds will be the product of the ratio of the distances traversed by the inverse ratio of the time-intervals occupied.

Let A and B be the two particles which move at a uniform rate; and let the respective distances traversed by them have the ratio of V to T, but let the time-intervals be as S to R. Then I say the speed of A will bear to the speed of

Fig. 44

B a ratio which is the product of the ratio of the distance V to the distance T and the time-interval R to the time-interval S.

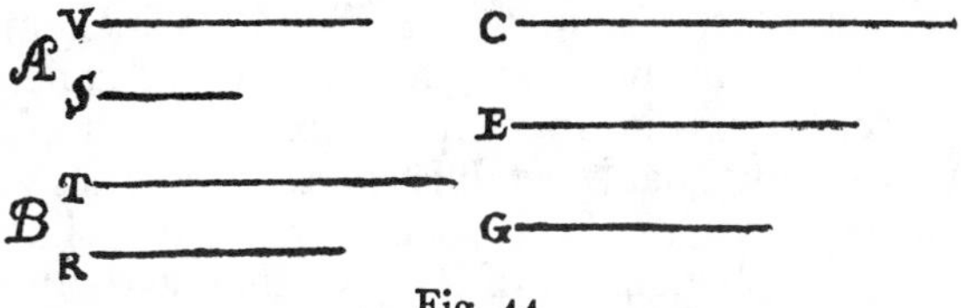

Let C be the speed at which A traverses the distance V during the time-interval S; and let the speed C bear the same ratio to another speed E as V bears to T; then E will be the speed at which B traverses the distance T during the time-interval S. If now the speed E is to another speed G as the time-interval R is to the time-interval S, then G will be the speed at which the

particle

particle B traverses the distance T during the time-interval R. Thus we have the speed C at which the particle A covers the distance V during the time S and also the speed G at which the particle B traverses the distance T during the time R. The ratio of C to G is the product of the ratio C to E and E to G; the ratio of C to E is by definition the same as the ratio of the distance V to distance T; and the ratio of E to G is the same as the ratio of R to S. Hence follows the proposition.

SALV. The preceding is what our Author has written concerning uniform motion. We pass now to a new and more discriminating consideration of naturally accelerated motion, such as that generally experienced by heavy falling bodies; following is the title and introduction.

[197]
NATURALLY ACCELERATED MOTION

The properties belonging to uniform motion have been discussed in the preceding section; but accelerated motion remains to be considered.

And first of all it seems desirable to find and explain a definition best fitting natural phenomena. For anyone may invent an arbitrary type of motion and discuss its properties; thus, for instance, some have imagined helices and conchoids as described by certain motions which are not met with in nature, and have very commendably established the properties which these curves possess in virtue of their definitions; but we have decided to consider the phenomena of bodies falling with an acceleration such as actually occurs in nature and to make this definition of accelerated motion exhibit the essential features of observed accelerated motions. And this, at last, after repeated efforts we trust we have succeeded in doing. In this belief we are confirmed mainly by the consideration that experimental results are seen to agree with and exactly correspond with those properties which have been, one after another, demonstrated by us. Finally, in the investigation of naturally accelerated motion we were led, by hand as it were, in following the habit and custom of

nature

nature herself, in all her various other processes, to employ only those means which are most common, simple and easy.

For I think no one believes that swimming or flying can be accomplished in a manner simpler or easier than that instinctively employed by fishes and birds.

When, therefore, I observe a stone initially at rest falling from an elevated position and continually acquiring new increments of speed, why should I not believe that such increases take place in a manner which is exceedingly simple and rather obvious to everybody? If now we examine the matter carefully we find no addition or increment more simple than that which repeats itself always in the same manner. This we readily understand when we consider the intimate relationship between time and motion; for just as uniformity of motion is defined by and conceived through equal times and equal spaces (thus we call a motion uniform when equal distances are traversed during equal time-intervals), so also we may, in a similar manner, through equal time-intervals, conceive additions of speed as taking place without complication; thus we may picture to our

[198]

mind a motion as uniformly and continuously accelerated when, during any equal intervals of time whatever, equal increments of speed are given to it. Thus if any equal intervals of time whatever have elapsed, counting from the time at which the moving body left its position of rest and began to descend, the amount of speed acquired during the first two time-intervals will be double that acquired during the first time-interval alone; so the amount added during three of these time-intervals will be treble; and that in four, quadruple that of the first time-interval. To put the matter more clearly, if a body were to continue its motion with the same speed which it had acquired during the first time-interval and were to retain this same uniform speed, then its motion would be twice as slow as that which it would have if its velocity had been acquired during *two* time-intervals.

And thus, it seems, we shall not be far wrong if we put the increment of speed as proportional to the increment of time; hence

hence the definition of motion which we are about to discuss may be stated as follows: A motion is said to be uniformly accelerated, when starting from rest, it acquires, during equal time-intervals, equal increments of speed.

SAGR. Although I can offer no rational objection to this or indeed to any other definition, devised by any author whomsoever, since all definitions are arbitrary, I may nevertheless without offense be allowed to doubt whether such a definition as the above, established in an abstract manner, corresponds to and describes that kind of accelerated motion which we meet in nature in the case of freely falling bodies. And since the Author apparently maintains that the motion described in his definition is that of freely falling bodies, I would like to clear my mind of certain difficulties in order that I may later apply myself more earnestly to the propositions and their demonstrations.

SALV. It is well that you and Simplicio raise these difficulties. They are, I imagine, the same which occurred to me when I first saw this treatise, and which were removed either by discussion with the Author himself, or by turning the matter over in my own mind.

SAGR. When I think of a heavy body falling from rest, that is, starting with zero speed and gaining speed in proportion to the

[199]

time from the beginning of the motion; such a motion as would, for instance, in eight beats of the pulse acquire eight degrees of speed; having at the end of the fourth beat acquired four degrees; at the end of the second, two; at the end of the first, one: and since time is divisible without limit, it follows from all these considerations that if the earlier speed of a body is less than its present speed in a constant ratio, then there is no degree of speed however small (or, one may say, no degree of slowness however great) with which we may not find this body travelling after starting from infinite slowness, i. e., from rest. So that if that speed which it had at the end of the fourth beat was such that, if kept uniform, the body would traverse two miles in an hour, and if keeping the speed which it had at the end of the

second

second beat, it would traverse one mile an hour, we must infer that, as the instant of starting is more and more nearly approached, the body moves so slowly that, if it kept on moving at this rate, it would not traverse a mile in an hour, or in a day, or in a year or in a thousand years; indeed, it would not traverse a span in an even greater time; a phenomenon which baffles the imagination, while our senses show us that a heavy falling body suddenly acquires great speed.

SALV. This is one of the difficulties which I also at the beginning, experienced, but which I shortly afterwards removed; and the removal was effected by the very experiment which creates the difficulty for you. You say the experiment appears to show that immediately after a heavy body starts from rest it acquires a very considerable speed: and I say that the same experiment makes clear the fact that the initial motions of a falling body, no matter how heavy, are very slow and gentle. Place a heavy body upon a yielding material, and leave it there without any pressure except that owing to its own weight; it is clear that if one lifts this body a cubit or two and allows it to fall upon the same material, it will, with this impulse, exert a new and greater pressure than that caused by its mere weight; and this effect is brought about by the [weight of the] falling body together with the velocity acquired during the fall, an effect which will be greater and greater according to the height of the fall, that is according as the velocity of the falling body becomes greater. From the quality and intensity of the blow we are thus enabled to accurately estimate the speed of a falling body. But tell me, gentlemen, is it not true that if a block be allowed to fall upon a stake from a height of four cubits and drives it into the earth,

[200]

say, four finger-breadths, that coming from a height of two cubits it will drive the stake a much less distance, and from the height of one cubit a still less distance; and finally if the block be lifted only one finger-breadth how much more will it accomplish than if merely laid on top of the stake without percussion? Certainly very little. If it be lifted only the thickness of a leaf, the effect will be altogether imperceptible. And since the
effect

effect of the blow depends upon the velocity of this striking body, can any one doubt the motion is very slow and the speed more than small whenever the effect [of the blow] is imperceptible? See now the power of truth; the same experiment which at first glance seemed to show one thing, when more carefully examined, assures us of the contrary.

But without depending upon the above experiment, which is doubtless very conclusive, it seems to me that it ought not to be difficult to establish such a fact by reasoning alone. Imagine a heavy stone held in the air at rest; the support is removed and the stone set free; then since it is heavier than the air it begins to fall, and not with uniform motion but slowly at the beginning and with a continuously accelerated motion. Now since velocity can be increased and diminished without limit, what reason is there to believe that such a moving body starting with infinite slowness, that is, from rest, immediately acquires a speed of ten degrees rather than one of four, or of two, or of one, or of a half, or of a hundredth; or, indeed, of any of the infinite number of small values [of speed]? Pray listen. I hardly think you will refuse to grant that the gain of speed of the stone falling from rest follows the same sequence as the diminution and loss of this same speed when, by some impelling force, the stone is thrown to its former elevation: but even if you do not grant this, I do not see how you can doubt that the ascending stone, diminishing in speed, must before coming to rest pass through every possible degree of slowness.

SIMP. But if the number of degrees of greater and greater slowness is limitless, they will never be all exhausted, therefore such an ascending heavy body will never reach rest, but will continue to move without limit always at a slower rate; but this is not the observed fact.

SALV. This would happen, Simplicio, if the moving body were to maintain its speed for any length of time at each degree of velocity; but it merely passes each point without delaying more than an instant: and since each time-interval however

[201]

small may be divided into an infinite number of instants, these

will

will always be sufficient [in number] to correspond to the infinite degrees of diminished velocity.

That such a heavy rising body does not remain for any length of time at any given degree of velocity is evident from the following: because if, some time-interval having been assigned, the body moves with the same speed in the last as in the first instant of that time-interval, it could from this second degree of elevation be in like manner raised through an equal height, just as it was transferred from the first elevation to the second, and by the same reasoning would pass from the second to the third and would finally continue in uniform motion forever.

SAGR. From these considerations it appears to me that we may obtain a proper solution of the problem discussed by philosophers, namely, what causes the acceleration in the natural motion of heavy bodies? Since, as it seems to me, the force [*virtù*] impressed by the agent projecting the body upwards diminishes continuously, this force, so long as it was greater than the contrary force of gravitation, impelled the body upwards; when the two are in equilibrium the body ceases to rise and passes through the state of rest in which the impressed impetus [*impeto*] is not destroyed, but only its excess over the weight of the body has been consumed—the excess which caused the body to rise. Then as the diminution of the outside impetus [*impeto*] continues, and gravitation gains the upper hand, the fall begins, but slowly at first on account of the opposing impetus [*virtù impressa*], a large portion of which still remains in the body; but as this continues to diminish it also continues to be more and more overcome by gravity, hence the continuous acceleration of motion.

SIMP. The idea is clever, yet more subtle than sound; for even if the argument were conclusive, it would explain only the case in which a natural motion is preceded by a violent motion, in which there still remains active a portion of the external force [*virtù esterna*]; but where there is no such remaining portion and the body starts from an antecedent state of rest, the cogency of the whole argument fails.

SAGR. I believe that you are mistaken and that this distinction
tion

tion between cases which you make is superfluous or rather nonexistent. But, tell me, cannot a projectile receive from the projector either a large or a small force [*virtù*] such as will throw it to a height of a hundred cubits, and even twenty or four or one?

[202]

SIMP. Undoubtedly, yes.

SAGR. So therefore this impressed force [*virtù impressa*] may exceed the resistance of gravity so slightly as to raise it only a finger-breadth; and finally the force [*virtù*] of the projector may be just large enough to exactly balance the resistance of gravity so that the body is not lifted at all but merely sustained. When one holds a stone in his hand does he do anything but give it a force impelling [*virtù impellente*] it upwards equal to the power [*facoltà*] of gravity drawing it downwards? And do you not continuously impress this force [*virtù*] upon the stone as long as you hold it in the hand? Does it perhaps diminish with the time during which one holds the stone?

And what does it matter whether this support which prevents the stone from falling is furnished by one's hand or by a table or by a rope from which it hangs? Certainly nothing at all. You must conclude, therefore, Simplicio, that it makes no difference whatever whether the fall of the stone is preceded by a period of rest which is long, short, or instantaneous provided only the fall does not take place so long as the stone is acted upon by a force [*virtù*] opposed to its weight and sufficient to hold it at rest.

SALV. The present does not seem to be the proper time to investigate the cause of the acceleration of natural motion concerning which various opinions have been expressed by various philosophers, some explaining it by attraction to the center, others to repulsion between the very small parts of the body, while still others attribute it to a certain stress in the surrounding medium which closes in behind the falling body and drives it from one of its positions to another. Now, all these fantasies, and others too, ought to be examined; but it is not really worth while. At present it is the purpose of our Author merely to

investigate

investigate and to demonstrate some of the properties of accelerated motion (whatever the cause of this acceleration may
be)—meaning thereby a motion, such that the momentum of its
velocity [*i momenti della sua velocità*] goes on increasing after
departure from rest, in simple proportionality to the time, which
is the same as saying that in equal time-intervals the body
receives equal increments of velocity; and if we find the properties [of accelerated motion] which will be demonstrated later are
realized in freely falling and accelerated bodies, we may conclude
that the assumed definition includes such a motion of falling
bodies and that their speed [*accelerazione*] goes on increasing as
the time and the duration of the motion.

[203]

SAGR. So far as I see at present, the definition might have
been put a little more clearly perhaps without changing the
fundamental idea, namely, uniformly accelerated motion is such
that its speed increases in proportion to the space traversed; so
that, for example, the speed acquired by a body in falling four
cubits would be double that acquired in falling two cubits and
this latter speed would be double that acquired in the first cubit.
Because there is no doubt but that a heavy body falling from
the height of six cubits has, and strikes with, a momentum
[*impeto*] double that it had at the end of three cubits, triple that
which it had at the end of one.

SALV. It is very comforting to me to have had such a companion in error; and moreover let me tell you that your proposition seems so highly probable that our Author himself admitted,
when I advanced this opinion to him, that he had for some time
shared the same fallacy. But what most surprised me was to
see two propositions so inherently probable that they com-
manded the assent of everyone to whom they were presented,
proven in a few simple words to be not only false, but impossible.

SIMP. I am one of those who accept the proposition, and
believe that a falling body acquires force [*vires*] in its descent, its
velocity increasing in proportion to the space, and that the
momentum [*momento*] of the falling body is doubled when it falls
from

from a doubled height; these propositions, it appears to me, ought to be conceded without hesitation or controversy.

SALV. And yet they are as false and impossible as that motion should be completed instantaneously; and here is a very clear demonstration of it. If the velocities are in proportion to the spaces traversed, or to be traversed, then these spaces are traversed in equal intervals of time; if, therefore, the velocity with which the falling body traverses a space of eight feet were double that with which it covered the first four feet (just as the one distance is double the other) then the time-intervals required for these passages would be equal. But for one and the same body to fall eight feet and four feet in the same time is possible only in the case of instantaneous [discontinuous] motion;

[204]

but observation shows us that the motion of a falling body occupies time, and less of it in covering a distance of four feet than of eight feet; therefore it is not true that its velocity increases in proportion to the space.

The falsity of the other proposition may be shown with equal clearness. For if we consider a single striking body the difference of momentum in its blows can depend only upon difference of velocity; for if the striking body falling from a double height were to deliver a blow of double momentum, it would be necessary for this body to strike with a doubled velocity; but with this doubled speed it would traverse a doubled space in the same time-interval; observation however shows that the time required for fall from the greater height is longer.

SAGR. You present these recondite matters with too much evidence and ease; this great facility makes them less appreciated than they would be had they been presented in a more abstruse manner. For, in my opinion, people esteem more lightly that knowledge which they acquire with so little labor than that acquired through long and obscure discussion.

SALV. If those who demonstrate with brevity and clearness the fallacy of many popular beliefs were treated with contempt instead of gratitude the injury would be quite bearable; but on the other hand it is very unpleasant and annoying to see men, who

who claim to be peers of anyone in a certain field of study, take for granted certain conclusions which later are quickly and easily shown by another to be false. I do not describe such a feeling as one of envy, which usually degenerates into hatred and anger against those who discover such fallacies; I would call it a strong desire to maintain old errors, rather than accept newly discovered truths. This desire at times induces them to unite against these truths, although at heart believing in them, merely for the purpose of lowering the esteem in which certain others are held by the unthinking crowd. Indeed, I have heard from our Academician many such fallacies held as true but easily refutable; some of these I have in mind.

SAGR. You must not withhold them from us, but, at the proper time, tell us about them even though an extra session be necessary. But now, continuing the thread of our talk, it would

[205]

seem that up to the present we have established the definition of uniformly accelerated motion which is expressed as follows:

A motion is said to be equally or uniformly accelerated when, starting from rest, its momentum (*celeritatis momenta*) receives equal increments in equal times.

SALV. This definition established, the Author makes a single assumption, namely,

The speeds acquired by one and the same body moving down planes of different inclinations are equal when the heights of these planes are equal.

By the height of an inclined plane we mean the perpendicular let fall from the upper end of the plane upon the horizontal line drawn through the lower end of the same plane. Thus, to illustrate, let the line AB be horizontal, and let the planes CA and CD be inclined to it; then the Author calls the perpendicular CB the "height" of the planes CA and CD; he supposes that the speeds acquired by one and the same body, descending along the planes CA and CD to the terminal points A and D are equal since the heights of these planes are the same, CB; and also it must be understood that this speed is that which would be acquired by the same body falling from C to B.

Sagr.

SAGR. Your assumption appears to me so reasonable that it ought to be conceded without question, provided of course there are no chance or outside resistances, and that the planes are hard and smooth, and that the figure of the moving body is perfectly round, so that neither plane nor moving body is rough. All resistance and opposition having been removed, my reason tells me at once that a heavy and perfectly round ball descending along the lines CA, CD, CB would reach the terminal points A, D, B, with equal momenta [*impeti eguali*].

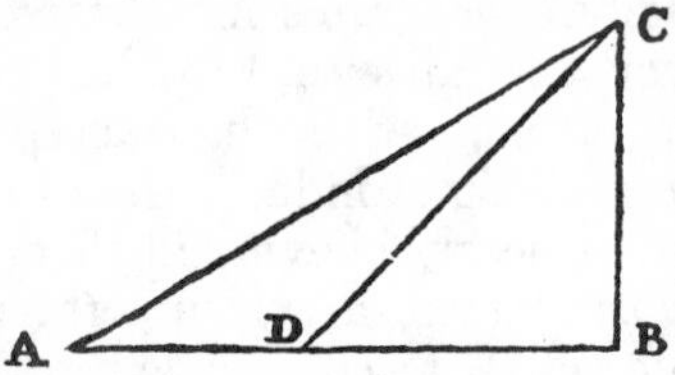

Fig. 45

SALV. Your words are very plausible; but I hope by experiment to increase the probability to an extent which shall be little short of a rigid demonstration.

[206]

Imagine this page to represent a vertical wall, with a nail driven into it; and from the nail let there be suspended a lead bullet of one or two ounces by means of a fine vertical thread, AB, say from four to six feet long, on this wall draw a horizontal line DC, at right angles to the vertical thread AB, which hangs about two finger-breadths in front of the wall. Now bring the thread AB with the attached ball into the position AC and set it free; first it will be observed to descend along the arc CBD, to pass the point B, and to travel along the arc BD, till it almost reaches the horizontal CD, a slight shortage being caused by the resistance of the air and the string; from this we may rightly infer that the ball in its descent through the arc CB acquired a momentum [*impeto*] on reaching B, which was just sufficient to carry it through a similar arc BD to the same height. Having repeated this experiment many times, let us now drive a nail into the wall close to the perpendicular AB, say at E or F, so that it projects out some five or six finger-breadths in order that the thread, again carrying the bullet through the arc CB, may strike upon the nail E when the bullet reaches B, and thus compel it to traverse the arc BG, described about E as center. From this

we

we can see what can be done by the same momentum [*impeto*] which previously starting at the same point B carried the same body through the arc BD to the horizontal CD. Now, gentlemen, you will observe with pleasure that the ball swings to the point G in the horizontal, and you would see the same thing happen if the obstacle were placed at some lower point, say at F, about which the ball would describe the arc BI, the rise of the

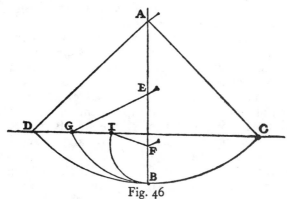

Fig. 46

ball always terminating exactly on the line CD. But when the nail is placed so low that the remainder of the thread below it will not reach to the height CD (which would happen if the nail were placed nearer B than to the intersection of AB with the
[207]
horizontal CD) then the thread leaps over the nail and twists itself about it.

This experiment leaves no room for doubt as to the truth of our supposition; for since the two arcs CB and DB are equal and similarly placed, the momentum [*momento*] acquired by the fall through the arc CB is the same as that gained by fall through the arc DB; but the momentum [*momento*] acquired at B, owing to fall through CB, is able to lift the same body [*mobile*] through the arc BD; therefore, the momentum acquired in the fall BD is equal to that which lifts the same body through the same arc from B to D; so, in general, every momentum acquired by fall through

through an arc is equal to that which can lift the same body through the same arc. But all these momenta [*momenti*] which cause a rise through the arcs BD, BG, and BI are equal, since they are produced by the same momentum, gained by fall through CB, as experiment shows. Therefore all the momenta gained by fall through the arcs DB, GB, IB are equal.

SAGR. The argument seems to me so conclusive and the experiment so well adapted to establish the hypothesis that we may, indeed, consider it as demonstrated.

SALV. I do not wish, Sagredo, that we trouble ourselves too much about this matter, since we are going to apply this principle mainly in motions which occur on plane surfaces, and not upon curved, along which acceleration varies in a manner greatly different from that which we have assumed for planes.

So that, although the above experiment shows us that the descent of the moving body through the arc CB confers upon it momentum [*momento*] just sufficient to carry it to the same height through any of the arcs BD, BG, BI, we are not able, by similar means, to show that the event would be identical in the case of a perfectly round ball descending along planes whose inclinations are respectively the same as the chords of these arcs. It seems likely, on the other hand, that, since these planes form angles at the point B, they will present an obstacle to the ball which has descended along the chord CB, and starts to rise along the chord BD, BG, BI.

In striking these planes some of its momentum [*impeto*] will be lost and it will not be able to rise to the height of the line CD; but this obstacle, which interferes with the experiment, once removed, it is clear that the momentum [*impeto*] (which gains
[208]
in strength with descent) will be able to carry the body to the same height. Let us then, for the present, take this as a postulate, the absolute truth of which will be established when we find that the inferences from it correspond to and agree perfectly with experiment. The author having assumed this single principle passes next to the propositions which he clearly demonstrates; the first of these is as follows:

THEOREM I, PROPOSITION I

The time in which any space is traversed by a body start-ing from rest and uniformly accelerated is equal to the time in which that same space would be traversed by the same body moving at a uniform speed whose value is the mean of the highest speed and the speed just before acceleration began.

Let us represent by the line AB the time in which the space CD is traversed by a body which starts from rest at C and is uniformly accelerated; let the final and highest value of the speed gained during the interval AB be represented by the line EB drawn at right angles to AB; draw the line AE, then all lines drawn from equidistant points on AB and parallel to BE will represent the increasing values of the speed, beginning with the instant A. Let the point F bisect the line EB; draw FG parallel to BA, and GA parallel to FB, thus forming a parallel-ogram AGFB which will be equal in area to the triangle AEB, since the side GF bisects the side AE at the point I; for if the parallel lines in the triangle AEB are extended to GI, then the sum of all the parallels contained in the quadrilateral is equal to the sum of those contained in the tri-angle AEB; for those in the triangle IEF are equal to those contained in the triangle GIA, while those included in the trapezium AIFB are common. Since each and every instant of time in the time-interval AB has its corresponding point on the line AB, from which points par-allels drawn in and limited by the triangle AEB represent the increasing values of the growing velocity, and since parallels contained within the rectangle rep-resent the values of a speed which is not increasing, but constant, it appears, in like manner, that the momenta [*momenta*] assumed by the moving body may also be represented, in the case of the accelerated motion, by the increasing parallels of the triangle AEB,

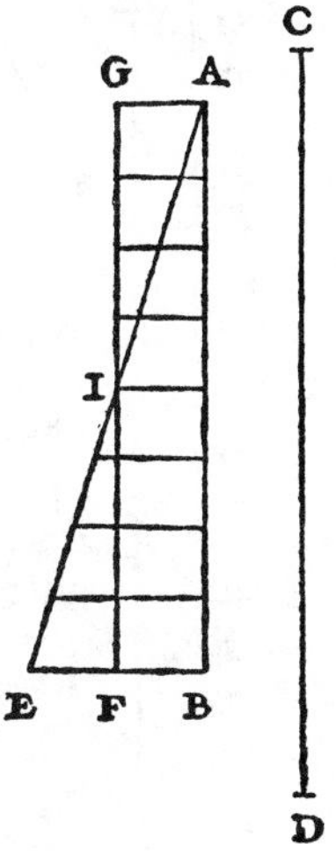

Fig. 47

[209]
AEB, and, in the case of the uniform motion, by the parallels of
the rectangle GB. For, what the momenta may lack in the first
part of the accelerated motion (the deficiency of the momenta
being represented by the parallels of the triangle AGI) is made
up by the momenta represented by the parallels of the triangle
IEF.

Hence it is clear that equal spaces will be traversed in equal
times by two bodies, one of which, starting from rest, moves with
a uniform acceleration, while the momentum of
the other, moving with uniform speed, is one-half
its maximum momentum under accelerated motion.   Q. E. D.

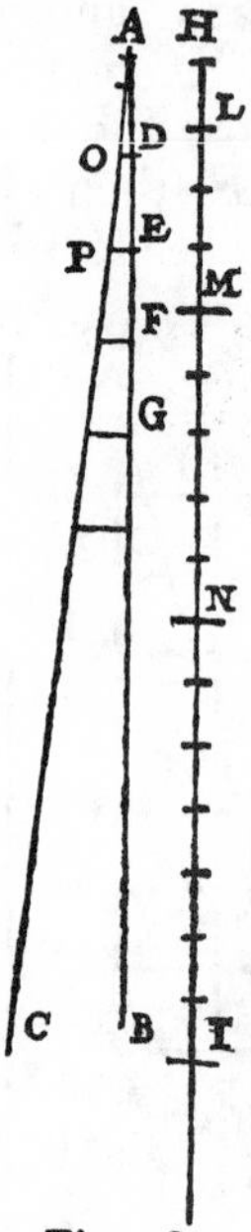

THEOREM II, PROPOSITION II

The spaces described by a body falling from rest
with a uniformly accelerated motion are to each
other as the squares of the time-intervals employed in traversing these distances.

Let the time beginning with any instant A be represented by the straight line AB in which are taken
any two time-intervals AD and AE. Let HI represent the distance through which the body, starting
from rest at H, falls with uniform acceleration. If
HL represents the space traversed during the time-
interval AD, and HM that covered during the interval AE, then the space MH stands to the space
LH in a ratio which is the square of the ratio of the
time AE to the time AD; or we may say simply that
the distances HM and HL are related as the squares
of AE and AD.

Fig. 48

Draw the line AC making any angle whatever with the line
AB; and from the points D and E, draw the parallel lines DO
and EP; of these two lines, DO represents the greatest velocity
attained during the interval AD, while EP represents the maximum velocity acquired during the interval AE. But it has
just been proved that so far as distances traversed are concerned

cerned it is precisely the same whether a body falls from rest with a uniform acceleration or whether it falls during an equal time-interval with a constant speed which is one-half the maximum speed attained during the accelerated motion. It follows therefore that the distances HM and HL are the same as would be traversed, during the time-intervals AE and AD, by uniform velocities equal to one-half those represented by DO and EP respectively. If, therefore, one can show that the distances HM and HL are in the same ratio as the squares of the time-intervals AE and AD, our proposition will be proven.

[210]

But in the fourth proposition of the first book [p. 157 above] it has been shown that the spaces traversed by two particles in uniform motion bear to one another a ratio which is equal to the product of the ratio of the velocities by the ratio of the times. But in this case the ratio of the velocities is the same as the ratio of the time-intervals (for the ratio of AE to AD is the same as that of ½ EP to ½ DO or of EP to DO). Hence the ratio of the spaces traversed is the same as the squared ratio of the time-intervals.   Q. E. D.

Evidently then the ratio of the distances is the square of the ratio of the final velocities, that is, of the lines EP and DO, since these are to each other as AE to AD.

COROLLARY I

Hence it is clear that if we take any equal intervals of time whatever, counting from the beginning of the motion, such as AD, DE, EF, FG, in which the spaces HL, LM, MN, NI are traversed, these spaces will bear to one another the same ratio as the series of odd numbers, 1, 3, 5, 7; for this is the ratio of the differences of the squares of the lines [which represent time], differences which exceed one another by equal amounts, this excess being equal to the smallest line [viz. the one representing a single time-interval]: or we may say [that this is the ratio] of the differences of the squares of the natural numbers beginning with unity.

While,

While, therefore, during equal intervals of time the velocities increase as the natural numbers, the increments in the distances traversed during these equal time-intervals are to one another as the odd numbers beginning with unity.

SAGR. Please suspend the discussion for a moment since there just occurs to me an idea which I want to illustrate by means of a diagram in order that it may be clearer both to you and to me.

Let the line AI represent the lapse of time measured from the initial instant A; through A draw the straight line AF making any angle whatever; join the terminal points I and F; divide the time AI in half at C; draw CB parallel to IF. Let us consider CB as the maximum value of the velocity which increases from zero at the beginning, in simple proportionality to the intercepts on the triangle ABC of lines drawn parallel to BC; or what is the same thing, let us suppose the velocity to increase in proportion to the time; then I admit without question, in view of the preceding argument, that the space described by a body falling in the aforesaid manner will be equal to the space traversed by the same body during the same length of time travelling with a uniform speed equal to EC, the half of BC. Further let us imagine that the body has fallen with accelerated motion so that, at the instant C, it has the velocity BC. It is clear that if the body continued to descend with the same speed BC, without acceleration, it would in the next time-interval CI traverse double the distance covered during the interval AC, with the uniform speed EC which is half of BC; but since the falling body acquires equal increments of speed during equal increments of time, it follows that the velocity BC, during the next time-interval

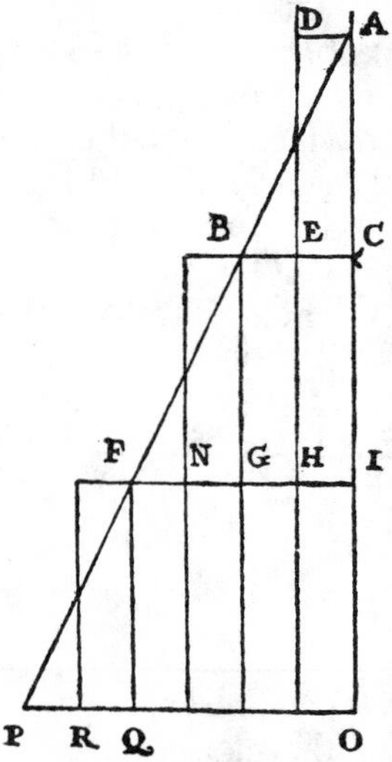

Fig. 49

[211]

interval CI will be increased by an amount represented by the parallels of the triangle BFG which is equal to the triangle ABC. If, then, one adds to the velocity GI half of the velocity FG, the highest speed acquired by the accelerated motion and determined by the parallels of the triangle BFG, he will have the uniform velocity with which the same space would have been described in the time CI; and since this speed IN is three times as great as EC it follows that the space described during the interval CI is three times as great as that described during the interval AC. Let us imagine the motion extended over another equal time-interval IO, and the triangle extended to APO; it is then evident that if the motion continues during the interval IO, at the constant rate IF acquired by acceleration during the time AI, the space traversed during the interval IO will be four times that traversed during the first interval AC, because the speed IF is four times the speed EC. But if we enlarge our triangle so as to include FPQ which is equal to ABC, still assuming the acceleration to be constant, we shall add to the uniform speed an increment RQ, equal to EC; then the value of the equivalent uniform speed during the time-interval IO will be five times that during the first time-interval AC; therefore the space traversed will be quintuple that during the first interval AC. It is thus evident by simple computation that a moving body starting from rest and acquiring velocity at a rate proportional to the time, will, during equal intervals of time, traverse distances which are related to each other as the odd numbers beginning with unity, 1, 3, 5; * or considering the total space traversed, that covered

[212[

in double time will be quadruple that covered during unit time; in triple time, the space is nine times as great as in unit time.

---

* As illustrating the greater elegance and brevity of modern analytical methods, one may obtain the result of Prop. II directly from the fundamental equation

$$s = \frac{1}{2} g \left(t^2_2 - t^2_1\right) = \frac{g}{2} \left(t_2 + t_1\right) \left(t_2 - t_1\right)$$

where g is the acceleration of gravity and s, the space traversed between the instants t_1 and t_2. If now $t_2 - t_1 = 1$, say one second, then $s = \frac{g}{2} \left(t_2 + t_1\right)$ where $t_2 + t_1$, must always be an odd number, seeing that it is the sum of two consecutive terms in the series of natural numbers. [*Trans.*]

And in general the spaces traversed are in the duplicate ratio of the times, i. e., in the ratio of the squares of the times.

SIMP. In truth, I find more pleasure in this simple and clear argument of Sagredo than in the Author's demonstration which to me appears rather obscure; so that I am convinced that matters are as described, once having accepted the definition of uniformly accelerated motion. But as to whether this acceleration is that which one meets in nature in the case of falling bodies, I am still doubtful; and it seems to me, not only for my own sake but also for all those who think as I do, that this would be the proper moment to introduce one of those experiments—and there are many of them, I understand—which illustrate in several ways the conclusions reached.

SALV. The request which you, as a man of science, make, is a very reasonable one; for this is the custom—and properly so—in those sciences where mathematical demonstrations are applied to natural phenomena, as is seen in the case of perspective, astronomy, mechanics, music, and others where the principles, once established by well-chosen experiments, become the foundations of the entire superstructure. I hope therefore it will not appear to be a waste of time if we discuss at considerable length this first and most fundamental question upon which hinge numerous consequences of which we have in this book only a small number, placed there by the Author, who has done so much to open a pathway hitherto closed to minds of speculative turn. So far as experiments go they have not been neglected by the Author; and often, in his company, I have attempted in the following manner to assure myself that the acceleration actually experienced by falling bodies is that above described.

A piece of wooden moulding or scantling, about 12 cubits long, half a cubit wide, and three finger-breadths thick, was taken; on its edge was cut a channel a little more than one finger in breadth; having made this groove very straight, smooth, and polished, and having lined it with parchment, also as smooth and polished as possible, we rolled along it a hard, smooth, and very round bronze ball. Having placed this

[213]

board

board in a sloping position, by lifting one end some one or two cubits above the other, we rolled the ball, as I was just saying, along the channel, noting, in a manner presently to be described, the time required to make the descent. We repeated this experiment more than once in order to measure the time with an accuracy such that the deviation between two observations never exceeded one-tenth of a pulse-beat. Having performed this operation and having assured ourselves of its reliability, we now rolled the ball only one-quarter the length of the channel; and having measured the time of its descent, we found it precisely one-half of the former. Next we tried other distances, comparing the time for the whole length with that for the half, or with that for two-thirds, or three-fourths, or indeed for any fraction; in such experiments, repeated a full hundred times, we always found that the spaces traversed were to each other as the squares of the times, and this was true for all inclinations of the plane, i. e., of the channel, along which we rolled the ball. We also observed that the times of descent, for various inclinations of the plane, bore to one another precisely that ratio which, as we shall see later, the Author had predicted and demonstrated for them.

For the measurement of time, we employed a large vessel of water placed in an elevated position; to the bottom of this vessel was soldered a pipe of small diameter giving a thin jet of water, which we collected in a small glass during the time of each descent, whether for the whole length of the channel or for a part of its length; the water thus collected was weighed, after each descent, on a very accurate balance; the differences and ratios of these weights gave us the differences and ratios of the times, and this with such accuracy that although the operation was repeated many, many times, there was no appreciable discrepancy in the results.

SIMP. I would like to have been present at these experiments; but feeling confidence in the care with which you performed them, and in the fidelity with which you relate them, I am satisfied and accept them as true and valid

SALV. Then we can proceed without discussion.

[214]

COROLLARY II

Secondly, it follows that, starting from any initial point, if we take any two distances, traversed in any time-intervals whatsoever, these time-intervals bear to one another the same ratio as one of the distances to the mean proportional of the two distances.

For if we take two distances ST and SY measured from the initial point S, the mean proportional of which is SX, the time of fall through ST is to the time of fall through SY as ST is to SX; or one may say the time of fall through SY is to the time of fall through ST as SY is to SX. Now since it has been shown that the spaces traversed are in the same ratio as the squares of the times; and since, moreover, the ratio of the space SY to the space ST is the square of the ratio SY to SX, it follows that the ratio of the times of fall through SY and ST is the ratio of the respective distances SY and SX.

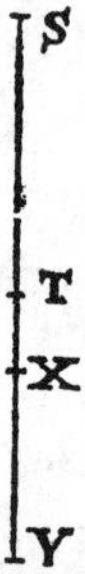

Fig. 50

SCHOLIUM

The above corollary has been proven for the case of vertical fall; but it holds also for planes inclined at any angle; for it is to be assumed that along these planes the velocity increases in the same ratio, that is, in proportion to the time, or, if you prefer, as the series of natural numbers.*

Salv. Here, Sagredo, I should like, if it be not too tedious to Simplicio, to interrupt for a moment the present discussion in order to make some additions on the basis of what has already been proved and of what mechanical principles we have already learned from our Academician. This addition I make for the better establishment on logical and experimental grounds, of the principle which we have above considered; and what is more important, for the purpose of deriving it geometrically, after first demonstrating a single lemma which is fundamental in the science of motion [impeti].

* The dialogue which intervenes between this Scholium and the following theorem was elaborated by Viviani, at the suggestion of Galileo. See *National Edition*, viii, 23. [*Trans.*]

SAGR. If the advance which you propose to make is such as will confirm and fully establish these sciences of motion, I will gladly devote to it any length of time. Indeed, I shall not only
[215]
be glad to have you proceed, but I beg of you at once to satisfy the curiosity which you have awakened in me concerning your proposition; and I think that Simplicio is of the same mind.

SIMP. Quite right.

SALV. Since then I have your permission, let us first of all consider this notable fact, that the momenta or speeds [*i momenti o le velocità*] of one and the same moving body vary with the inclination of the plane.

The speed reaches a maximum along a vertical direction, and for other directions diminishes as the plane diverges from the vertical. Therefore the impetus, ability, energy, [*l'impeto, il talento, l'energia*] or, one might say, the momentum [*il momento*] of descent of the moving body is diminished by the plane upon which it is supported and along which it rolls.

For the sake of greater clearness erect the line AB perpendicular to the horizontal AC; next draw AD, AE, AF, etc., at different inclinations to the horizontal. Then I say that all the momentum of the falling body is along the vertical and is a maximum when it falls in that direction; the momentum is less along DA and still less along EA, and even less yet along the more inclined plane FA.

Finally on the horizontal plane the momentum vanishes altogether; the body finds itself in a condition of indifference as to motion or rest; has no inherent tendency to move in any direction, and offers no resistance to being set in motion. For just as a heavy body or system of bodies cannot of itself move upwards, or recede from the common center [*comun centro*] toward which all heavy things tend, so it is impossible for any body of its own accord to assume any motion other than

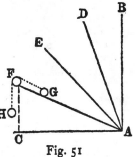

Fig. 51

one which carries it nearer to the aforesaid common center. Hence, along the horizontal, by which we understand a surface, every point of which is equidistant from this same common center, the body will have no momentum whatever.

This

[216]

This change of momentum being clear, it is here necessary for me to explain something which our Academician wrote when in Padua, embodying it in a treatise on mechanics prepared solely for the use of his students, and proving it at length and conclusively when considering the origin and nature of that marvellous machine, the screw. What he proved is the manner in which the momentum [*impeto*] varies with the inclination of the plane, as for instance that of the plane FA, one end of which is elevated through a vertical distance FC. This direction FC is that along which the momentum of a heavy body becomes a maximum; let us discover what ratio this momentum bears to that of the same body moving along the inclined plane FA. This ratio, I say, is the inverse of that of the aforesaid lengths. Such is the lemma preceding the theorem which I hope to demonstrate a little later.

It is clear that the impelling force [*impeto*] acting on a body in descent is equal to the resistance or least force [*resistenza o forza minima*] sufficient to hold it at rest. In order to measure this force and resistance [*forza e resistenza*] I propose to use the weight of another body. Let us place upon the plane FA a body G connected to the weight H by means of a cord passing over the point F; then the body H will ascend or descend, along the perpendicular, the same distance which the body G ascends or descends along the inclined plane FA; but this distance will not be equal to the rise or fall of G along the vertical in which direction alone G, as other bodies, exerts its force [*resistenza*]. This is clear. For if we consider the motion of the body G, from A to F, in the triangle AFC to be made up of a horizontal component AC and a vertical component CF, and remember that this body experiences no resistance to motion along the horizontal (because by such a

[217]

motion the body neither gains nor loses distance from the common center of heavy things) it follows that resistance is met only in consequence of the body rising through the vertical distance CF. Since then the body G in moving from A to F offers resistance only in so far as it rises through the vertical distance CF, while the other body H must fall vertically through the entire distance FA, and since this ratio is maintained whether the motion be large or small, the two bodies being inextensibly connected, we are able to assert positively that, in case of equilibrium (bodies at rest) the

momenta,

momenta, the velocities, or their tendency to motion [*propensioni al moto*], i. e., the spaces which would be traversed by them in equal times, must be in the inverse ratio to their weights. This is what has been demonstrated in every case of mechanical motion.* So that, in order to hold the weight G at rest, one must give H a weight smaller in the same ratio as the distance CF is smaller than FA. If we do this, FA:FC=weight G:weight H; then equilibrium will occur, that is, the weights H and G will have the same impelling forces [*momenti eguali*], and the two bodies will come to rest.

And since we are agreed that the impetus, energy, momentum or tendency to motion of a moving body is as great as the force or least resistance [*forza o resistenza minima*] sufficient to stop it, and since we have found that the weight H is capable of preventing motion in the weight G, it follows that the less weight H whose entire force [*momento totale*] is along the perpendicular, FC, will be an exact measure of the component of force [*momento parziale*] which the larger weight G exerts along the plane FA. But the measure of the total force [*total momento*] on the body G is its own weight, since to prevent its fall it is only necessary to balance it with an equal weight, provided this second weight be free to move vertically; therefore the component of the force [*momento parziale*] on G along the inclined plane FA will bear to the maximum and total force on this same body G along the perpendicular FC the same ratio as the weight H to the weight G. This ratio is, by construction, the same which the height, FC, of the inclined plane bears to the length FA. We have here the lemma which I proposed to demonstrate and which, as you will see, has been assumed by our Author in the second part of the sixth proposition of the present treatise.

SAGR. From what you have shown thus far, it appears to me that one might infer, arguing *ex aequali con la proportione perturbata*, that the tendencies [*momenti*] of one and the same body to move along planes differently inclined, but having the same vertical height, as FA and FI, are to each other inversely as the lengths of the planes.

[218]

SALV. Perfectly right. This point established, I pass to the demonstration of the following theorem:

* A near approach to the principle of virtual work enunciated by John Bernoulli in 1717. [*Trans.*]

If a body falls freely along smooth planes inclined at any angle whatsoever, but of the same height, the speeds with which it reaches the bottom are the same.

First we must recall the fact that on a plane of any inclination whatever a body starting from rest gains speed or momentum [*la quantità dell'impeto*] in direct proportion to the time, in agreement with the definition of naturally accelerated motion given by the Author. Hence, as he has shown in the preceding proposition, the distances traversed are proportional to the squares of the times and therefore to the squares of the speeds. The speed relations are here the same as in the motion first studied [i. e., *vertical motion*], since in each case the gain of speed is proportional to the time.

Let AB be an inclined plane whose height above the level BC is AC. As we have seen above the force impelling [*l'impeto*] a body

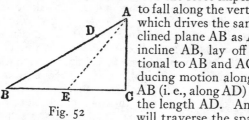

Fig. 52

to fall along the vertical AC is to the force which drives the same body along the inclined plane AB as AB is to AC. On the incline AB, lay off AD a third proportional to AB and AC; then the force producing motion along AC is to that along AB (i. e., along AD) as the length AC is to the length AD. And therefore the body will traverse the space AD, along the incline AB, in the same time which it would occupy in falling the vertical distance AC, (since the forces [*momenti*] are in the same ratio as these distances); also the speed at C is to the speed at D as the distance AC is to the distance AD. But, according to the definition of accelerated motion, the speed at B is to the speed of the same body at D as the time required to traverse AB is to the time required for AD; and, according to the last corollary of the second proposition, the time of passing through the distance AB bears to the time of passing through AD the same ratio as the distance AC (a mean proportional between AB and AD) to AD. Accordingly the two speeds at B and C each bear to the speed at D the same ratio, namely, that of the distances AC and AD; hence they are equal. This is the theorem which I set out to prove.

From the above we are better able to demonstrate the following third proposition of the Author in which he employs the following principle, namely, the time required to traverse an inclined plane

is

is to that required to fall through the vertical height of the plane
in the same ratio as the length of the plane to its height.

[219]

For, according to the second corollary of the second proposition,
if BA represents the time required to pass over the distance BA,
the time required to pass the distance AD will be a mean proportional between these two distances and will be represented by
the line AC; but if AC represents the time needed to traverse AD
it will also represent the time required to fall through the distance
AC, since the distances AC and AD are traversed in equal times;
consequently if AB represents the time required for AB then AC
will represent the time required for AC. Hence the times required
to traverse AB and AC are to each other as the distances AB and
AC.

In like manner it can be shown that the time required to fall
through AC is to the time required for any other incline AE as
the length AC is to the length AE; therefore, *ex aequali*, the time of
fall along the incline AB is to that along AE as the distance AB is
to the distance AE, etc.*

One might by application of this same theorem, as Sagredo will
readily see, immediately demonstrate the sixth proposition of the
Author; but let us here end this digression which Sagredo has
perhaps found rather tedious, though I consider it quite important
for the theory of motion.

SAGR. On the contrary it has given me great satisfaction, and
indeed I find it necessary for a complete grasp of this principle.

SALV. I will now resume the reading of the text.

[215]

THEOREM III, PROPOSITION III

If one and the same body, starting from rest, falls along
an inclined plane and also along a vertical, each having the
same height, the times of descent will be to each other as
the lengths of the inclined plane and the vertical.

Let AC be the inclined plane and AB the perpendicular, each
having the same vertical height above the horizontal, namely,
BA; then I say, the time of descent of one and the same body

---

* Putting this argument in a modern and evident notation, one has
$AC = \frac{1}{2} g t_c^2$ and $AD = \frac{1}{2} \frac{AC}{AB} g t_d^2$ If now $\overline{AC}^2 = AB . AD$, it follows at
once that $t_d = t_c$. [*Trans.*]   Q. D. E.

[216]

along the plane AC bears a ratio to the time of fall along the
perpendicular AB, which is the same as the ratio of the length
AC to the length AB. Let DG, EI and LF be any lines parallel
to the horizontal CB; then it follows from
what has preceded that a body starting from
A will acquire the same speed at the point G
as at D, since in each case the vertical fall is
the same; in like manner the speeds at I and
E will be the same; so also those at L and F.
And in general the speeds at the two extremities of any parallel drawn from any point on
AB to the corresponding point on AC will be
equal.

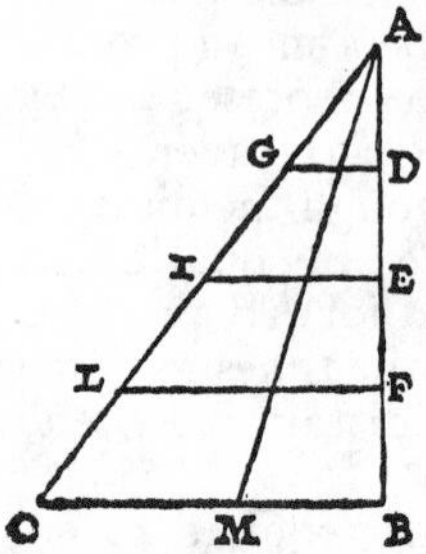

Fig. 53   Thus the two distances AC and AB are
traversed at the same speed. But it has already been proved

[217]

that if two distances are traversed by a body moving with equal
speeds, then the ratio of the times of descent will be the ratio of
the distances themselves; therefore, the time of descent along
AC is to that along AB as the length of the plane AC is to the
vertical distance AB.   Q. E. D.

[218]

SAGR. It seems to me that the above could have been proved
clearly and briefly on the basis of a proposition already demonstrated, namely, that the distance traversed in the case of
accelerated motion along AC or AB is the same as that covered

[219]

by a uniform speed whose value is one-half the maximum speed,
CB; the two distances AC and AB having been traversed at the
same uniform speed it is evident, from Proposition I, that the
times of descent will be to each other as the distances.

COROLLARY

Hence we may infer that the times of descent along planes
having different inclinations, but the same vertical height stand

to

to one another in the same ratio as the lengths of the planes. For consider any plane AM extending from A to the horizontal CB; then it may be demonstrated in the same manner that the time of descent along AM is to the time along AB as the distance AM is to AB; but since the time along AB is to that along AC as the length AB is to the length AC, it follows, *ex æquali*, that as AM is to AC so is the time along AM to the time along AC.

THEOREM IV, PROPOSITION IV

The times of descent along planes of the same length but of different inclinations are to each other in the inverse ratio of the square roots of their heights

From a single point B draw the planes BA and BC, having the same length but different inclinations; let AE and CD be horizontal lines drawn to meet the perpendicular BD; and

[220]

let BE represent the height of the plane AB, and BD the height of BC; also let BI be a mean proportional to BD and BE; then the ratio of BD to BI is equal to the square root of the ratio of BD to BE. Now, I say, the ratio of the times of descent along BA and BC is the ratio of BD to BI; so that the time of descent along BA is related to the height of the other plane BC, namely BD as the time along BC is related to the height BI. Now it must be proved that the time of descent along BA is to that along BC as the length BD is to the length BI.

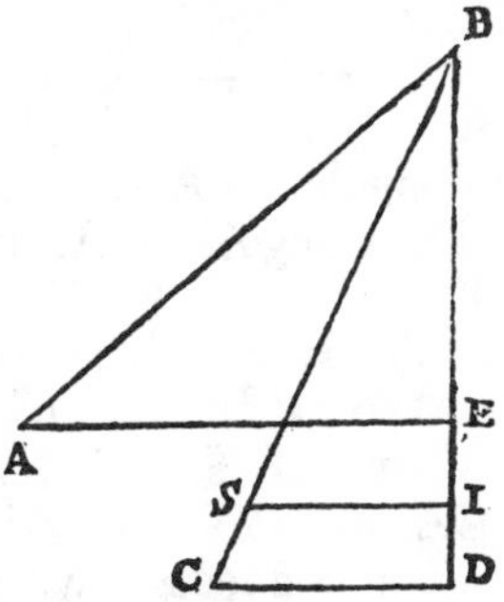

Draw IS parallel to DC; and since it has been shown that the time of fall along BA is to that along the vertical BE as BA is to BE; and also that the time along BE is to that along BD as BE is to BI; and likewise that the time along BD is to that along BC as BD is to BC, or as BI to BS; it follows, *ex æquali*, that the time along BA is to that along BC as BA to BS, or BC to BS. However, BC is to BS as BD is to BI; hence follows our proposition.

Fig. 54

THEOREM V, PROPOSITION V

The times of descent along planes of different length, slope and height bear to one another a ratio which is equal to the product of the ratio of the lengths by the square root of the inverse ratio of their heights.

Draw the planes AB and AC, having different inclinations, lengths, and heights. My theorem then is that the ratio of the time of descent along AC to that along AB is equal to the product of the ratio of AC to AB by the square root of the inverse ratio of their heights.

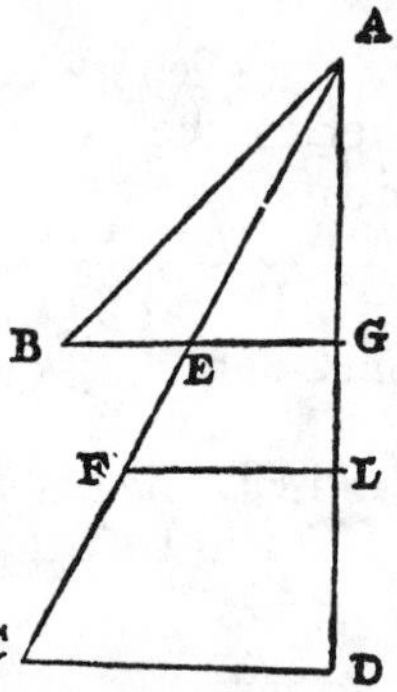

For let AD be a perpendicular to which are drawn the horizontal lines BG and CD; also let AL be a mean proportional to the heights AG and AD; from the point L draw a horizontal line meeting AC in F; accordingly AF will be a mean proportional between AC and AE. Now since the time of descent along AC is to that along AE as the length AF is to AE; and since the time along AE is to that along AB as AE is to AB, it is clear that the time along AC is to that along AB as AF is to AB.

Fig. 55

[221]

Thus it remains to be shown that the ratio of AF to AB is equal to the product of the ratio of AC to AB by the ratio of AG to AL, which is the inverse ratio of the square roots of the heights DA and GA. Now it is evident that, if we consider the line AC in connection with AF and AB, the ratio of AF to AC is the same as that of AL to AD, or AG to AL which is the square root of the ratio of the heights AG and AD; but the ratio of AC to AB is the ratio of the lengths themselves. Hence follows the theorem.

THEOREM VI, PROPOSITION VI

If from the highest or lowest point in a vertical circle there be drawn any inclined planes meeting the circumference the times

times of descent along these chords are each equal to the other.

On the horizontal line GH construct a vertical circle. From its lowest point—the point of tangency with the horizontal—draw the diameter FA and from the highest point, A, draw inclined planes to B and C, any points whatever on the circumference; then the times of descent along these are equal. Draw BD and CE perpendicular to the diameter; make AI a mean proportional between the heights of the planes, AE and AD; and since the rectangles FA.AE and FA. AD are respectively equal to the squares of AC and AB, while the rectangle FA.AE is to the rectangle FA.AD as AE is to AD, it follows that the square of AC

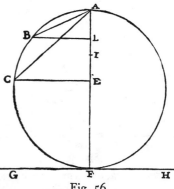

Fig. 56

is to the square of AB as the length AE is to the length AD. But since the length AE is to AD as the square of AI is to the square of AD, it follows that the squares on the lines AC and AB are to each other as the squares on the lines AI and AD, and hence also the length AC is to the length AB as AI is to AD. But it has previously been demonstrated that the ratio of the time of descent along AC to that along AB is equal to the product of the two ratios AC to AB and AD to AI; but this last ratio is the same as that of AB to AC. Therefore the ratio of the time of descent along AC to that along AB is the product of the two ratios, AC to AB and AB to AC. The ratio of these times is therefore unity. Hence follows our proposition.

By use of the principles of mechanics [*ex mechanicis*] one may obtain the same result, namely, that a falling body will require equal times to traverse the distances CA and DA, indicated in the following figure. Lay off BA equal to DA, and let fall the

[222]

perpendiculars BE and DF; it follows from the principles of mechanics

mechanics that the component of the momentum [*momentum ponderis*] acting along the inclined plane ABC is to the total momentum [i. e., the momentum of the body falling freely] as

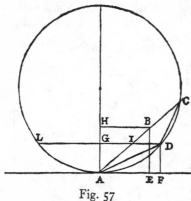

Fig. 57

BE is to BA; in like manner the momentum along the plane AD is to its total momentum [i. e., the momentum of the body falling freely] as DF is to DA, or to BA. Therefore the momentum of this same weight along the plane DA is to that along the plane ABC as the length DF is to the length BE; for this reason, this same weight will in equal times according to the second proposition of the first book, traverse spaces along the planes CA and DA which are to each other as the lengths BE and DF. But it can be shown that CA is to DA as BE is to DF. Hence the falling body will traverse the two paths CA and DA in equal times.

Moreover the fact that CA is to DA as BE is to DF may be demonstrated as follows: Join C and D; through D, draw the line DGL parallel to AF and cutting the line AC in I; through B draw the line BH, also parallel to AF. Then the angle ADI will be equal to the angle DCA, since they subtend equal arcs LA and DA, and since the angle DAC is common, the sides of the triangles, CAD and DAI, about the common angle will be proportional to each other; accordingly as CA is to DA so is DA to IA, that is as BA is to IA, or as HA is to GA, that is as BE is to DF.   E. D.

The same proposition may be more easily demonstrated as follows: On the horizontal line AB draw a circle whose diameter DC is vertical. From the upper end of this diameter draw any inclined plane, DF, extending to meet the circumference; then, I say, a body will occupy the same time in falling along the plane DF as along the diameter DC. For draw FG parallel

to

to AB and perpendicular to DC; join FC; and since the time of
fall along DC is to that along DG as the mean proportional

[223]

between CD and GD is to GD itself; and since also DF is a
mean proportional between DC and DG, the angle DFC inscribed in a semicircle being a right-
angle, and FG being perpendicular
to DC, it follows that the time of
fall along DC is to that along DG as
the length FD is to GD. But it has
already been demonstrated that the
time of descent along DF is to that
along DG as the length DF is to DG;
hence the times of descent along DF
and DC each bear to the time of fall
along DG the same ratio; consequently they are equal.

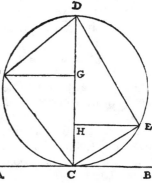

Fig. 58

In like manner it may be shown
that if one draws the chord CE from
the lower end of the diameter, also the line EH parallel to the
horizon, and joins the points E and D, the time of descent along
EC,will be the same as that along the diameter, DC.

COROLLARY I

From this it follows that the times of descent along all chords
drawn through either C or D are equal one to another.

COROLLARY II

It also follows that, if from any one point there be drawn a
vertical line and an inclined one along which the time of descent
is the same, the inclined line will be a chord of a semicircle of
which the vertical line is the diameter.

COROLLARY III

Moreover the times of descent along inclined planes will be
equal when the vertical heights of equal lengths of these planes
are

are to each other as the lengths of the planes themselves; thus it is clear that the times of descent along CA and DA, in the figure just before the last, are equal, provided the vertical height of AB (AB being equal to AD), namely, BE, is to the vertical height DF as CA is to DA.

SAGR. Please allow me to interrupt the lecture for a moment in order that I may clear up an idea which just occurs to me; one which, if it involve no fallacy, suggests at least a freakish and

[224]

interesting circumstance, such as often occurs in nature and in the realm of necessary consequences.

If, from any point fixed in a horizontal plane, straight lines be drawn extending indefinitely in all directions, and if we imagine a point to move along each of these lines with constant speed, all starting from the fixed point at the same instant and moving with equal speeds, then it is clear that all of these moving points will lie upon the circumference of a circle which grows larger and larger, always having the aforesaid fixed point as its center; this circle spreads out in precisely the same manner as the little waves do in the case of a pebble allowed to drop into quiet water, where the impact of the stone starts the motion in all directions, while the point of impact remains the center of these ever-expanding circular waves. But imagine a vertical plane from the highest point of which are drawn lines inclined at every angle and extending indefinitely; imagine also that heavy particles descend along these lines each with a naturally accelerated motion and each with a speed appropriate to the inclination of its line. If these moving particles are always visible, what will be the locus of their positions at any instant? Now the answer to this question surprises me, for I am led by the preceding theorems to believe that these particles will always lie upon the circumference of a single circle, ever increasing in size as the particles recede farther and farther from the point at which their motion began. To be more definite, let A be the fixed point from which are drawn the lines AF and AH inclined at any angle whatsoever. On the perpendicular AB take any two points C and D about which, as centers, circles are described

passing

passing through the point A, and cutting the inclined lines at the points F, H, B, E, G, I. From the preceding theorems it is clear that, if particles start, at the same instant, from A and descend along these lines, when one is at E another will be at G and another at I; at a later instant they will be found simultaneously at F, H and B; these, and indeed an infinite number of other particles [225] travelling along an infinite number of different slopes will at successive instants always lie upon a single ever-expanding circle. The two kinds of motion occurring in nature give rise therefore to two infinite series of circles, at once resembling and

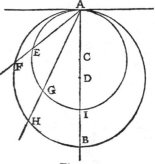

Fig. 59

differing from each other; the one takes its rise in the center of an infinite number of concentric circles; the other has its origin in the contact, at their highest points, of an infinite number of eccentric circles; the former are produced by motions which are equal and uniform; the latter by motions which are neither uniform nor equal among themselves, but which vary from one to another according to the slope.

Further, if from the two points chosen as origins of motion, we draw lines not only along horizontal and vertical planes but in all directions then just as in the former cases, beginning at a single point ever-expanding circles are produced, so in the latter case an infinite number of spheres are produced about a single point, or rather a single sphere which expands in size without limit; and this in two ways, one with the origin at the center, the other on the surface of the spheres.

Salv. The idea is really beautiful and worthy of the clever mind of Sagredo.

Simp. As for me, I understand in a general way how the two kinds of natural motions give rise to the circles and spheres; and yet as to the production of circles by accelerated motion and its proof, I am not entirely clear; but the fact that one can take the

the origin of motion either at the inmost center or at the very top of the sphere leads one to think that there may be some great mystery hidden in these true and wonderful results, a mystery related to the creation of the universe (which is said to be spherical in shape), and related also to the seat of the first cause [*prima causa*].

SALV. I have no hesitation in agreeing with you. But profound considerations of this kind belong to a higher science than ours [*a più alte dottrine che le nostre*]. We must be satisfied to belong to that class of less worthy workmen who procure from the quarry the marble out of which, later, the gifted sculptor produces those masterpieces which lay hidden in this rough and shapeless exterior. Now, if you please, let us proceed.

[226]

THEOREM VII, PROPOSITION VII

If the heights of two inclined planes are to each other in the same ratio as the squares of their lengths, bodies starting from rest will traverse these planes in equal times.

Take two planes of different lengths and different inclinations, AE and AB, whose heights are AF and AD: let AF be to AD as

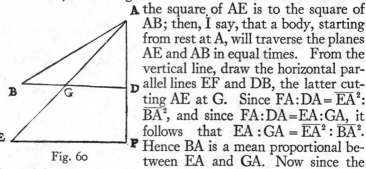

the square of AE is to the square of AB; then, I say, that a body, starting from rest at A, will traverse the planes AE and AB in equal times. From the vertical line, draw the horizontal parallel lines EF and DB, the latter cutting AE at G. Since $FA:DA = \overline{EA}^2 : \overline{BA}^2$, and since $FA:DA = EA:GA$, it follows that $EA:GA = \overline{EA}^2 : \overline{BA}^2$. Hence BA is a mean proportional between EA and GA. Now since the

Fig. 60

time of descent along AB bears to the time along AG the same ratio which AB bears to AG and since also the time of descent along AG is to the time along AE as AG is to a mean proportional between AG and AE, that is, to AB, it follows, *ex æquali*, that

that the time along AB is to the time along AE as AB is to itself.
Therefore the times are equal.   Q. E. D.

THEOREM VIII, PROPOSITION VIII

The times of descent along all inclined planes which intersect
one and the same vertical circle, either at its highest or
lowest point, are equal to the time of fall along the vertical
diameter; for those planes which fall short of this diameter
the times are shorter; for planes which cut this diameter, the
times are longer.

Let AB be the vertical diameter of a circle which touches the
horizontal plane. It has already
been proven that the times of descent along planes drawn from
either end, A or B, to the circumference are equal. In order
to show that the time of descent
[227]
along the plane DF which falls
short of the diameter is shorter
we may draw the plane DB which
is both longer and less steeply inclined than DF; whence it follows
that the time along DF is less than
that along DB and consequently

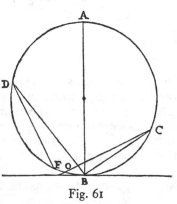

Fig. 61

along AB. In like manner, it is shown that the time of descent along CO which cuts the diameter is greater: for it is both
longer and less steeply inclined than CB. Hence follows the
theorem.

THEOREM IX, PROPOSITION IX

If from any point on a horizontal line two planes, inclined
at any angle, are drawn, and if they are cut by a line which
makes with them angles alternately equal to the angles between these planes and the horizontal, then the times required to traverse those portions of the plane cut off by
the aforesaid line are equal.

Through

Through the point C on the horizontal line X, draw two planes CD and CE inclined at any angle whatever: at any point in the line CD lay off the angle CDF equal to the angle XCE; let the line DF cut CE at F so that the angles CDF and CFD are alternately equal to XCE and LCD; then, I say, the

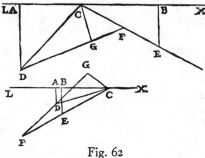

times of descent over CD and CF are equal. Now since the angle CDF is equal to the angle XCE by construction, it is evident that the angle CFD must be equal to the angle DCL. For if the common angle DCF be subtracted from the three angles of the tri-

Fig. 62   angle CDF, together equal

to two right angles, (to which are also equal all the angles which can be described about the point C on the lower side of the line LX) there remain in the triangle two angles, CDF and CFD, equal to the two angles XCE and LCD; but, by hypothesis, the angles CDF and XCE are equal; hence the remaining angle CFD is equal to the remainder DCL. Take CE equal to CD; from the points D and E draw DA and EB perpendicular to the horizontal line XL; and from the point C draw CG perpendicular to DF. Now since the angle CDG is equal to the angle ECB and since DGC and CBE are right angles, it follows that the triangles CDG and CBE are equiangular; consequently DC:CG=CE:EB. But DC is equal to CE, and therefore CG is equal to EB. Since also the angles at C and at A, in the triangle DAC, are equal to the angles at F and G in the triangle CGF, we have CD:DA = FC:CG and, *permutando*, DC:CF=DA:CG=DA:BE. Thus the ratio of the heights of the equal planes CD and CE is the same as the ratio of the lengths DC and CF. Therefore, by

[228]

Corollary I of Prop. VI, the times of descent along these planes will be equal.   Q. E. D.

An alternative proof is the following: Draw FS perpendicular to

to the horizontal line AS. Then, since the triangle CSF is similar to the triangle DGC, we have SF:FC=GC:CD; and since the triangle CFG is similar to the triangle DCA, we have FC:CG=CD:DA.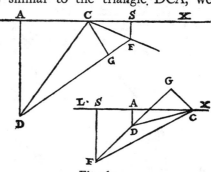
Hence, *ex æquali*, SF: CG=CG:DA. Therefore CG is a mean proportional between SF and DA, while DA:SF= $\overline{DA}^2:\overline{CG}^2$. Again since the triangle ACD is similar to the triangle CGF, we have DA:DC=GC: CF and, *permutando*, DA:CG = DC:CF: also

Fig. 63

$\overline{DA}^2:\overline{CG}^2=\overline{DC}^2:\overline{CF}^2$. But it has been shown that $\overline{DA}^2:\overline{CG}^2=$ DA:SF. Therefore $\overline{DC}^2:\overline{CF}^2=$DA:FS. Hence from the above Prop. VII, since the heights DA and FS of the planes CD and CF are to each other as the squares of the lengths of the planes, it follows that the times of descent along these planes will be equal.

THEOREM X, PROPOSITION X

The times of descent along inclined planes of the same height, but of different slope, are to each other as the lengths of these planes; and this is true whether the motion starts from rest or whether it is preceded by a fall from a constant height.

Let the paths of descent be along ABC and ABD to the horizontal plane DC so that the falls along BD and BC are preceded by the fall along AB; then, I say, that the time of descent along BD is to the time of descent along BC as the length BD is to BC. Draw the horizontal line AF and extend DB until it cuts this

[229]

line at F; let FE be a mean proportional between DF and FB; draw EO parallel to DC; then AO will be a mean proportional between CA and AB. If now we represent the time of fall along AB

AB by the length AB, then the time of descent along FB will be represented by the distance FB; so also the time of fall through the entire distance AC will be represented by the mean proportional AO: and for the entire distance FD by FE. Hence the time of fall along the remainder, BC, will be represented by

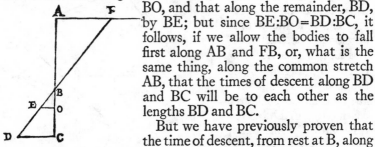

BO, and that along the remainder, BD, by BE; but since BE:BO=BD:BC, it follows, if we allow the bodies to fall first along AB and FB, or, what is the same thing, along the common stretch AB, that the times of descent along BD and BC will be to each other as the lengths BD and BC.

But we have previously proven that the time of descent, from rest at B, along BD is to the time along BC in the ratio which the length BD bears to BC. Hence the times of descent along different planes of constant height are to each other as the lengths of these planes, whether the motion starts from rest or is preceded by a fall from a constant height. Q. E. D.

Fig. 64

Theorem XI, Proposition XI

If a plane be divided into any two parts and if motion along it starts from rest, then the time of descent along the first part is to the time of descent along the remainder as the length of this first part is to the excess of a mean proportional between this first part and the entire length over this first part.

Let the fall take place, from rest at A, through the entire distance AB which is divided at any point C; also let AF be a mean proportional between the entire length BA and the first part AC; then CF will denote the excess of the mean proportional FA over the first part AC. Now, I say, the time of descent along AC will be to the time of subsequent fall through CB as the length AC is to CF. Fig. 65 This is evident, because the time along AC is to the time along the entire distance AB as AC is to the mean proportional AF.

Therefore,

Therefore, *dividendo*, the time along AC will be to the time along the remainder CB as AC is to CF. If we agree to represent the time along AC by the length AC then the time along CB will be represented by CF. Q. E. D.

[230]

In case the motion is not along the straight line ACB but along the broken line ACD to the horizontal line BD, and if from F we draw the horizontal line FE, it may in like manner be proved that the time along AC is to the time along the inclined line CD as AC is to CE. For the time along AC is to the time along CB as AC is to CF; but it has already been shown that the time along CB, after the fall through the distance AC, is to the time along CD, after descent through the same distance AC, as CB is to CD, or, as CF is to CE; therefore, *ex æquali*, the time along AC will be to the time along CD as the length AC is to the length CE.

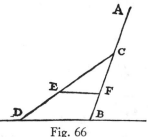

Fig. 66

THEOREM XII, PROPOSITION XII

If a vertical plane and any inclined plane are limited by two horizontals, and if we take mean proportionals between the lengths of these planes and those portions of them which lie between their point of intersection and the upper horizontal, then the time of fall along the perpendicular bears to the time required to traverse the upper part of the perpendicular plus the time required to traverse the lower part of the intersecting plane the same ratio which the entire length of the vertical bears to a length which is the sum of the mean proportional on the vertical plus the excess of the entire length of the inclined plane over its mean proportional.

Let AF and CD be two horizontal planes limiting the vertical plane AC and the inclined plane DF; let the two last-mentioned planes intersect at B. Let AR be a mean proportional between the

the entire vertical AC and its upper part AB; and let FS be a
mean proportional between FD and its upper part FB. Then,
I say, the time of fall along the entire vertical path AC bears to
the time of fall along its upper portion AB plus the time of fall

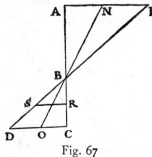

Fig. 67

along the lower part of the inclined
plane, namely, BD, the same ratio
which the length AC bears to the
mean proportional on the vertical,
namely, AR, plus the length SD which
is the excess of the entire plane DF
over its mean proportional FS.

Join the points R and S giving a
horizontal line RS. Now since the
time of fall through the entire distance AC is to the time along the

portion AB as CA is to the mean proportional AR it follows
that, if we agree to represent the time of fall through AC by
the distance AC, the time of fall through the distance AB will
be represented by AR; and the time of descent through the remainder, BC, will be represented by RC. But, if the time along
AC is taken to be equal to the length AC, then the time along
FD will be equal to the distance FD; and we may likewise infer
that the time of descent along BD, when preceded by a fall along
FB or AB, is numerically equal to the distance DS. Therefore

[231]

the time required to fall along the path AC is equal to AR plus
RC; while the time of descent along the broken line ABD will be
equal to AR plus SD.   Q. E. D.

The same thing is true if, in place of a vertical plane, one
takes any other plane, as for instance NO; the method of proof
is also the same.

PROBLEM I, PROPOSITION XIII

Given a perpendicular line of limited length, it is required
to find a plane having a vertical height equal to the given
perpendicular and so inclined that a body, having fallen
from rest along the perpendicular, will make its descent
along

along the inclined plane in the same time which it occupied in falling through the given perpendicular.

Let AB denote the given perpendicular: prolong this line to
C making BC equal to AB, and draw the horizontal lines CE
and AG. It is required to draw a plane from B to the horizontal
line CE such that after a body starting from rest at A has
fallen through the distance AB, it will complete its path along
this plane in an equal time. Lay off CD equal to BC, and draw
the line BD. Construct the line BE equal to the sum of BD and
DC; then, I say, BE is the required plane. Prolong EB till it
intersects the horizontal AG at G. Let GF be a mean proportional between GE and GB;
then EF:FB=EG:GF, and $\overline{EF}^2$:
$\overline{FB}^2=\overline{EG}^2:\overline{GF}^2=EG:GB$. But
EG is twice GB; hence the square
of EF is twice the square of FB;
so also is the square of DB twice
the square of BC. Consequently
EF:FB=DB:BC, and *componendo
et permutando*, EB:DB + BC =
BF:BC. But EB=DB + BC;

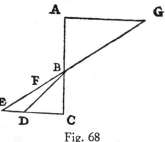

Fig. 68

hence BF=BC=BA. If we agree that the length AB shall represent the time of fall along the line AB, then GB will represent
the time of descent along GB, and GF the time along the entire
distance GE; therefore BF will represent the time of descent
along the difference of these paths, namely, BE, after fall from
G or from A.   Q. E. F.

[232]
PROBLEM II, PROPOSITION XIV

Given an inclined plane and a perpendicular passing
through it, to find a length on the upper part of the perpendicular through which a body will fall from rest in the
same time which is required to traverse the inclined plane
after fall through the vertical distance just determined.

Let AC be the inclined plane and DB the perpendicular. It is
required to find on the vertical AD a length which will be
traversed

traversed by a body, falling from rest, in the same time which is needed by the same body to traverse the plane AC after the aforesaid fall. Draw the horizontal CB; lay off AE such that BA + 2AC:AC=AC:AE, and lay off AR such that BA:AC= EA:AR. From R draw RX perpendicular to DB; then, I say, X is the point sought. For since BA + 2AC:AC=AC:AE, it follows, *dividendo*, that BA + AC:AC=CE:AE. And since BA:AC=EA:AR, we have, *componendo*, BA + AC:AC=ER: RA. But BA + AC:AC=CE:AE, hence CE:EA=ER:RA= sum of· the antecedents: sum of the consequents=CR:RE.

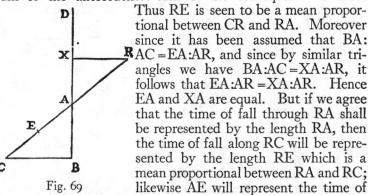

Thus RE is seen to be a mean proportional between CR and RA. Moreover since it has been assumed that BA: AC =EA:AR, and since by similar triangles we have BA:AC=XA:AR, it follows that EA:AR =XA:AR. Hence EA and XA are equal. But if we agree that the time of fall through RA shall be represented by the length RA, then the time of fall along RC will be represented by the length RE which is a mean proportional between RA and RC;

Fig. 69

likewise AE will represent the time of descent along AC after descent along RA or along AX. But the time of fall through XA is represented by the length XA, while RA represents the time through RA. But it has been shown that XA and AE are equal.   Q. E. F.

PROBLEM III, PROPOSITION XV

Given a vertical line and a plane inclined to it, it is required to find a length on the vertical line below its point of intersection which will be traversed in the same time as the inclined plane, each of these motions having been preceded by a fall through the given vertical line.

Let AB represent the vertical line and BC the inclined plane; it is required to find a length on the perpendicular below its point of intersection, which after a fall from A will be traversed in the same

same time which is needed for BC after an identical fall from A. Draw the horizontal AD, intersecting the prolongation of CB at D; let DE be a mean proportional between CD and DB; lay

[233]

off BF equal to BE; also let AG be a third proportional to BA and AF. Then, I say, BG is the distance which a body, after falling through AB, will traverse in the same time which is needed for the plane BC after the same preliminary fall. For if we assume that the time of fall along AB is represented by AB, then the time for DB will be represented by DB. And since DE is a mean proportional between BD and DC, this same DE will represent the time of descent along the entire distance DC while BE will represent the time required for the difference of these paths, namely, BC, provided in each case the fall is from rest at D or at A. In like manner we may infer that BF represents the time

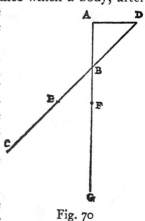

Fig. 70

of descent through the distance BG after the same preliminary fall; but BF is equal to BE. Hence the problem is solved.

THEOREM XIII, PROPOSITION XVI

If a limited inclined plane and a limited vertical line are drawn from the same point, and if the time required for a body, starting from rest, to traverse each of these is the same, then a body falling from any higher altitude will traverse the inclined plane in less time than is required for the vertical line.

Let EB be the vertical line and CE the inclined plane, both starting from the common point E, and both traversed in equal times by a body starting from rest at E; extend the vertical line upwards to any point A, from which falling bodies are allowed to start. Then, I say that, after the fall through AE, the inclined plane EC will be traversed in less time than the per-

pendicular

pendicular EB. Join CB, draw the horizontal AD, and prolong
CE backwards until it meets the latter in D; let DF be a mean
proportional between CD and DE while AG is made a mean
proportional between BA and AE. Draw FG and DG; then

[234]

since the times of descent along EC and EB, starting from rest
at E, are equal, it follows, according to Corollary II of Proposi-

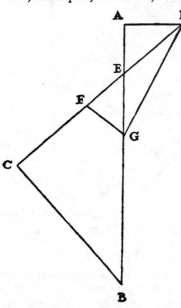

Fig. 71

tion VI that the angle at C is a
right angle; but the angle at A
is also a right angle and the
angles at the vertex E are
equal; hence the triangles AED
and CEB are equiangular and
the sides about the equal angles
are proportional; hence BE:
EC = DE:EA. Consequently
the rectangle BE.EA is equal
to the rectangle CE.ED; and
since the rectangle CD.DE exceeds the rectangle CE.ED by
the square of ED, and since the
rectangle BA.AE exceeds the
rectangle BE.EA by the square
of EA, it follows that the excess
of the rectangle CD.DE over
the rectangle BA.AE, or what
is the same thing, the excess of
the square of FD over the
square of AG, will be equal to

the excess of the square of DE over the square of AE, which excess is equal to the square of AD. Therefore $\overline{FD}^2 = \overline{GA}^2 +$
$\overline{AD}^2 = \overline{GD}^2$. Hence DF is equal to DG, and the angle DGF
is equal to the angle DFG while the angle EGF is less than the
angle EFG, and the opposite side EF is less than the opposite
side EG. If now we agree to represent the time of fall through
AE by the length AE, then the time along DE will be represented
by DE. And since AG is a mean proportional between BA and
AE,

AE, it follows that AG will represent the time of fall through the total distance AB, and the difference EG will represent the time of fall, from rest at A, through the difference of path EB.

In like manner EF represents the time of descent along EC, starting from rest at D or falling from rest at A. But it has been shown that EF is less than EG; hence follows the theorem.

COROLLARY

From this and the preceding proposition, it is clear that the vertical distance covered by a freely falling body, after a preliminary fall, and during the time-interval required to traverse an inclined plane, is greater than the length of the inclined plane, but less than the distance traversed on the inclined plane during an equal time, without any preliminary fall. For since we have just shown that bodies falling from an elevated point A will traverse the plane EC in Fig. 71 in a shorter time than the vertical EB, it is evident that the distance along EB which will be traversed during a time equal to that of descent along EC will be less than the whole of EB. But now in order to show that this vertical distance is greater than the length of the inclined plane EC, we reproduce Fig. 70 of the preceding theorem in which the vertical length BG is traversed in the same time as BC after a preliminary fall through AB. That BG is greater than BC is shown as follows: since BE and FB are equal [235] while BA is less than BD, it follows that FB will bear to BA a greater ratio than EB bears to BD; and, *componendo*, FA will bear to BA a greater ratio than ED to DB; but FA:AB = GF:FB (since AF is a mean proportional between BA and AG) and in like manner ED:BD = CE:EB. Hence GB bears to BF a greater ratio than CB bears to BE; therefore GB is greater than BC.

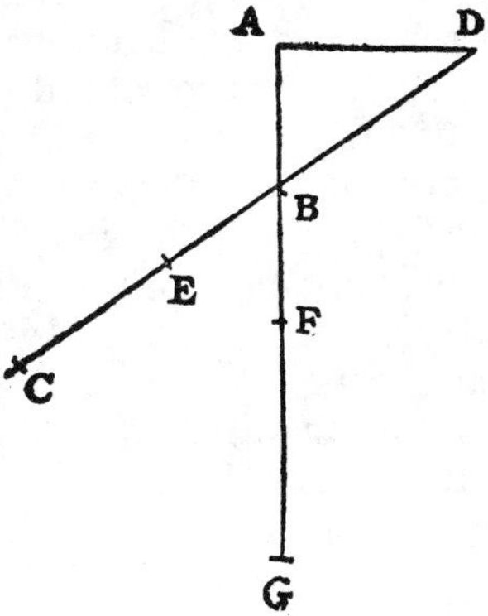

Fig. 72

PROBLEM IV, PROPOSITION XVII

Given a vertical line and an inclined plane, it is required to lay off a distance along the given plane which will be traversed by a body, after fall along the perpendicular, in the same time-interval which is needed for this body to fall from rest through the given perpendicular.

Let AB be the vertical line and BE the inclined plane. The problem is to determine on BE a distance such that a body, after falling through AB, will traverse it in a time equal to that required to traverse the perpendicular AB itself, starting from rest.

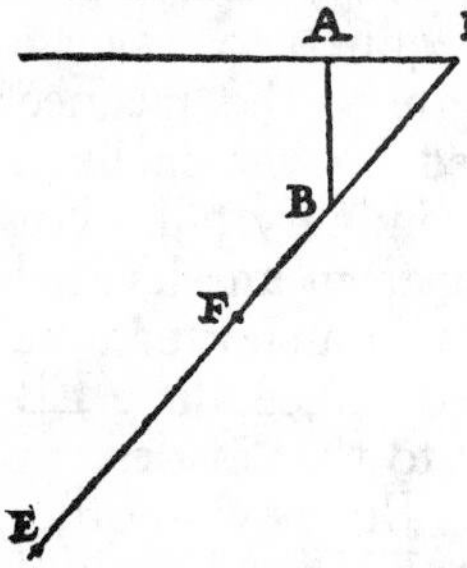

Draw the horizontal AD and extend the plane until it meets this line in D. Lay off FB equal to BA; and choose the point E such that BD:FD =DF:DE. Then, I say, the time of descent along BE, after fall through AB, is equal to the time of fall, from rest at A, through AB. For, if we assume that the length AB represents the time of fall through AB, then the time of fall through DB will be represented by the time DB; and since BD:FD =DF:DE, it follows that DF will represent the time of descent along the entire plane DE while BF represents the time through the portion BE starting from rest at D; but the time of descent along BE after the preliminary descent along DB is the same as that after a preliminary fall through AB. Hence the time of descent along BE after AB will be BF which of course is equal to the time of fall through AB from rest at A.'   Q. E. F.

Fig. 73

[236]

PROBLEM V, PROPOSITION XVIII

Given the distance through which a body will fall vertically from rest during a given time-interval, and given also a smaller time-interval, it is required to locate another [equal] vertical

vertical distance which the body will traverse during this given smaller time-interval.

Let the vertical line be drawn through A, and on this line lay off the distance AB which is traversed by a body falling from rest at A, during a time which may also be represented by AB. Draw the horizontal line CBE, and on it lay off BC to represent the given interval of time which is shorter than AB. It is required to locate, in the perpendicular above mentioned, a distance which is equal to AB and which will be described in a time equal to BC. Join the points A and C; then, since BC<BA, it follows that the angle BAC<angle BCA. Construct the angle CAE equal to BCA and let E be the point where AE intersects the horizontal line; draw ED at right angles to AE, cutting the vertical at D; lay off DF equal to BA. Then, I say, that FD is that portion of the vertical

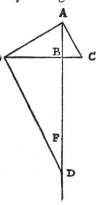

Fig. 74

which a body starting from rest at A will traverse during the assigned time-interval BC. For, if in the right-angled triangle AED a perpendicular be drawn from the right-angle at E to the opposite side AD, then AE will be a mean proportional between DA and AB while BE will be a mean proportional between BD and BA, or between FA and AB (seeing that FA is equal to DB); and since it has been agreed to represent the time of fall through AB by the distance AB, it follows that AE, or EC, will represent the time of fall through the entire distance AD, while EB will represent the time through AF. Consequently the remainder BC will represent the time of fall through the remaining distance FD. Q. E. F.

[237]
PROBLEM VI, PROPOSITION XIX

Given the distance through which a body falls in a vertical line from rest and given also the time of fall, it is required to find the time in which the same body will, later, traverse an an equal distance chosen anywhere in the same vertical line.

On the vertical line AB, lay off AC equal to the distance fallen from rest at A, also locate at random an equal distance DB.

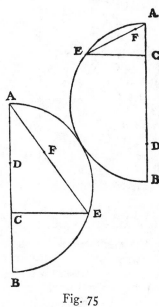

Let the time of fall through AC be represented by the length AC. It is required to find the time necessary to traverse DB after fall from rest at A. About the entire length AB describe the semicircle AEB; from C draw CE perpendicular to AB; join the points A and E; the line AE will be longer than EC; lay off EF equal to EC. Then, I say, the difference FA will represent the time required for fall through DB. For since AE is a mean proportional between BA and AC and since AC represents the time of fall through AC, it follows that AE will represent the time through the entire distance AB. And since CE is a mean proportional between DA and AC (seeing that DA = BC) it follows that CE, that is, EF, will represent the time of fall through AD. Hence the difference AF will represent the time of fall through the difference DB.   Q. E. D.

Fig. 75

COROLLARY

Hence it is inferred that if the time of fall from rest through any given distance is represented by that distance itself, then the time of fall, after the given distance has been increased by a certain amount, will be represented by the excess of the mean proportional between the increased distance and the original distance over the mean proportional between the original distance and the increment. Thus, for instance, if we agree that AB

AB represents the time of fall, from rest at A, through the distance AB, and that AS is the increment, the time required to traverse AB, after fall through SA, will be the excess of the mean proportional between SB and BA over the mean proportional between BA and AS.

[238]
Problem VII, Proposition XX

Given any distance whatever and a portion of it laid off from the point at which motion begins, it is required to find another portion which lies at the other end of the distance and which is traversed in the same time as the first given portion.

Fig. 76

Let the given distance be CB and let CD be that part of it which is laid off from the beginning of motion. It is required to find another part, at the end B, which is traversed in the same time as the assigned portion CD. Let BA be a mean proportional between BC and CD; also let CE be a third proportional to BC and CA. Then, I say, EB will be the distance which, after fall from C, will be traversed in the same time as CD itself. For if we agree that CB shall represent the time through the entire distance CB, then BA (which, of course, is a mean proportional between BC and CD) will represent the time along CD; and since CA is a mean proportional between BC and CE, it follows that CA will be the time through CE; but the total length CB represents the time through the total distance CB. Therefore the difference BA will be the time along the difference of distances, EB, after falling from C; but this same BA was the time of fall through CD. Consequently the distances CD and EB are traversed, from rest at A, in equal times.   Q. E. F.

Fig. 77

Theorem XIV, Proposition XXI

If, on the path of a body falling vertically from rest, one lays off a portion which is traversed in any time you please and

and whose upper terminus coincides with the point where the motion begins, and if this fall is followed by a motion deflected along any inclined plane, then the space traversed along the inclined plane, during a time-interval equal to that occupied in the previous vertical fall, will be greater than twice, and less than three times, the length of the vertical fall.

Let AB be a vertical line drawn downwards from the horizontal line AE, and let it represent the path of a body falling from rest at A; choose any portion AC of this path. Through C draw any inclined plane, CG, along which the motion is continued after fall through AC. Then, I say, that the distance

[239]

traversed along this plane CG, during the time-interval equal to that of the fall through AC, is more than twice, but less

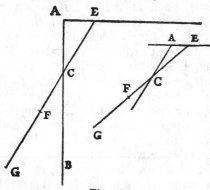

Fig. 78

than three times, this same distance AC. Let us lay off CF equal to AC, and extend the plane GC until it meets the horizontal in E; choose G such that CE: EF =EF:EG. If now we assume that the time of fall along AC is represented by the length AC, then CE will represent the time of descent along CE, while CF, or CA, will represent the time of descent along

CG. It now remains to be shown that the distance CG is more than twice, and less than three times, the distance CA itself. Since CE:EF =EF:EG, it follows that CE:EF =CF:FG; but EC<EF; therefore CF will be less than FG and GC will be more than twice FC, or AC. Again since FE<2EC (for EC is greater than CA, or CF), we have GF less than twice FC, and also GC less than three times CF, or CA.   Q. E. D.

This proposition may be stated in a more general form; since what

what has been proven for the case of a vertical and inclined plane holds equally well in the case of motion along a plane of any inclination followed by motion along any plane of less steepness, as can be seen from the adjoining figure. The method of proof is the same.

[240]
PROBLEM VIII, PROPOSITION XXII

Given two unequal time-intervals, also the distance through which a body will fall along a vertical line, from rest, during the shorter of these intervals, it is required to pass through the highest point of this vertical line a plane so inclined that the time of descent along it will be equal to the longer of the given intervals.

Let A represent the longer and B the shorter of the two unequal time-intervals, also let CD represent the length of the

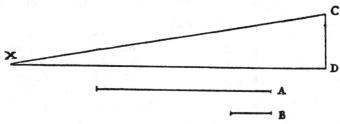

Fig. 79

vertical fall, from rest, during the time B. It is required to pass through the point C a plane of such a slope that it will be traversed in the time A.

Draw from the point C to the horizontal a line CX of such a length that B:A =CD:CX. It is clear that CX is the plane along which a body will descend in the given time A. For it has been shown that the time of descent along an inclined plane bears to the time of fall through its vertical height the same ratio which the length of the plane bears to its vertical height. Therefore the time along CX is to the time along CD as the length CX is to the length CD, that is, as the time-interval A is
to

to the time-interval B: but B is the time required to traverse the vertical distance, CD, starting from rest; therefore A is the time required for descent along the plane CX.

PROBLEM IX, PROPOSITION XXIII

Given the time employed by a body in falling through a certain distance along a vertical line, it is required to pass through the lower terminus of this vertical fall, a plane so inclined that this body will, after its vertical fall, traverse on this plane, during a time-interval equal to that of the vertical fall, a distance equal to any assigned distance, pro-

[241]

vided this assigned distance is more than twice and less than three times, the vertical fall.

Let AS be any vertical line, and let AC denote both the length of the vertical fall, from rest at A, and also the time

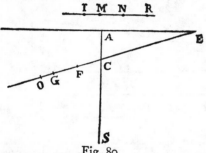

Fig. 80

required for this fall. Let IR be a distance more than twice and less than three times, AC. It is required to pass a plane through the point C so inclined that a body, after fall through AC, will, during the time AC, traverse a distance equal to IR. Lay off RN and NM each equal to AC. Through the point C, draw a plane CE meeting the horizontal, AE, at such a point that IM:MN =AC:CE. Extend the plane to O, and lay off CF, FG and GO equal to RN, NM, and MI respectively. Then, I say, the time along the inclined plane CO, after fall through AC, is equal to the time of fall, from rest at A, through AC. For since OG:GF =FC:CE, it follows, *componendo*, that OF:FG =OF:FC =FE:EC, and since an antecedent is to its consequent as the sum of the antecedents is to the sum of the consequents, we have OE:EF =EF:EC. Thus EF is a mean proportional between OE and EC. Having agreed to

represent

represent the time of fall through AC by the length AC it follows that EC will represent the time along EC, and EF the time along the entire distance EO, while the difference CF will represent the time along the difference CO; but CF = CA; therefore the problem is solved. For the time CA is the time of fall, from rest at A, through CA while CF (which is equal to CA) is the time required to traverse CO after descent along EC or after fall through AC.   Q. E. F.

It is to be remarked also that the same solution holds if the antecedent motion takes place, not along a vertical, but along an inclined plane. This case is illustrated in the following figure where the antecedent motion is along the inclined plane AS
[242]
underneath the horizontal AE. The proof is identical with the preceding.

SCHOLIUM

On careful attention, it will be clear that, the nearer the given line IR approaches to three times the length AC, the nearer the

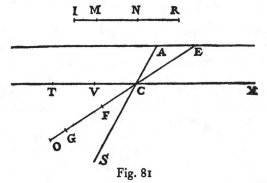

Fig. 81

inclined plane, CO, along which the second motion takes place, approaches the perpendicular along which the space traversed, during the time AC, will be three times the distance AC. For if IR be taken nearly equal to three times AC, then IM will be almost equal to MN; and since, by construction,
IM:

IM:MN =AC:CE, it follows that CE is but little greater than CA: consequently the point E will lie near the point A, and the lines CO and CS, forming a very acute angle, will almost coincide. But, on the other hand, if the given line, IR, be only the least bit longer than twice AC, the line IM will be very short; from which it follows that AC will be very small in comparison with CE which is now so long that it almost coincides with the horizontal line drawn through C. Hence we can infer that, if, after descent along the inclined plane AC of the adjoining figure, the motion is continued along a horizontal line, such as CT, the distance traversed by a body, during a time equal to the time of fall through AC, will be exactly twice the distance AC. The argument here employed is the same as the preceding. For it is clear, since OE:EF =EF:EC, that FC measures the time of descent along CO. But, if the horizontal line TC which is twice as long as CA, be divided into two equal parts at V then this line must be extended indefinitely in the direction of X before it will intersect the line AE produced; and accordingly the ratio of the infinite length TX to the infinite length VX is the same as the ratio of the infinite distance VX to the infinite distance CX.

The same result may be obtained by another method of approach, namely, by returning to the same line of argument which was employed in the proof of the first proposition. Let us

[243]

consider the triangle ABC, which, by lines drawn parallel to its base, represents for us a velocity increasing in proportion to the time; if these lines are infinite in number, just as the points in the line AC are infinite or as the number of instants in any interval of time is infinite, they will form the area of the triangle. Let us now suppose that the maximum velocity attained—that represented by the line BC—to be continued, without acceleration and at constant value through another interval of time equal to the first. From these velocities will be built up, in a similar manner, the area of the parallelogram ADBC, which is twice that of the triangle ABC; accordingly the distance traversed with these velocities during any given interval of time will be

twice

twice that traversed with the velocities represented by the triangle during an equal interval of time. But along a horizontal plane the motion is uniform since here it experiences neither acceleration nor retardation; therefore we con-clude that the distance CD traversed during a time-interval equal to AC is twice the distance AC; for the latter is covered by a motion, starting from rest and increasing in speed in proportion to the parallel lines in the triangle, while the former is traversed by a motion represented by the parallel lines of the parallelogram which, being also infinite in number, yield an area twice that of the triangle.

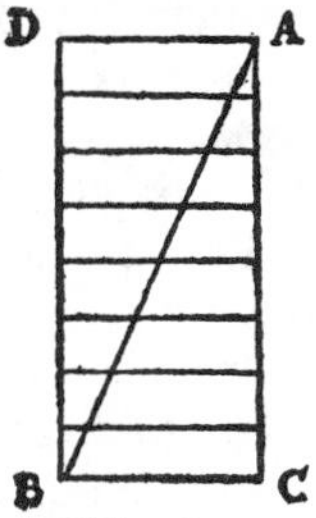

Fig. 82

Furthermore we may remark that any velocity once imparted to a moving body will be rigidly maintained as long as the external causes of acceleration or retardation are removed, a condition which is found only on horizontal planes; for in the case of planes which slope downwards there is already present a cause of acceleration, while on planes sloping upward there is retardation; from this it follows that motion along a horizontal plane is perpetual; for, if the velocity be uniform, it cannot be diminished or slackened, much less destroyed. Further, al-though any velocity which a body may have acquired through natural fall is permanently maintained so far as its own nature [*suapte natura*] is concerned, yet it must be remembered that if, after descent along a plane inclined downwards, the body is deflected to a plane inclined upward, there is already existing in this latter plane a cause of retardation; for in any such plane this same body is subject to a natural acceleration downwards. Accordingly we have here the superposition of two different states, namely, the velocity acquired during the preceding fall which if acting alone would carry the body at a uniform rate to infinity, and the velocity which results from a natural accelera-tion downwards common to all bodies. It seems altogether reasonable, therefore, if we wish to trace the future history of a body which has descended along some inclined plane and has been deflected along some plane inclined upwards, for us to

assume

assume that the maximum speed acquired during descent is permanently maintained during the ascent. In the ascent, however, there supervenes a natural inclination downwards, namely, a motion which, starting from rest, is accelerated at the

[244]

usual rate. If perhaps this discussion is a little obscure, the following figure will help to make it clearer.

Let us suppose that the descent has been made along the downward sloping plane AB, from which the body is deflected so as to continue its motion along the upward sloping plane BC; and first let these planes be of equal length and placed so as to make equal angles with the horizontal line GH. Now it is well known that a body, starting from rest at A, and descending along AB, acquires a speed which is proportional to the time,

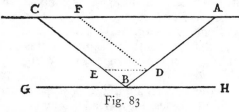

Fig. 83

which is a maximum at B, and which is maintained by the body so long as all causes of fresh acceleration or retardation are removed; the acceleration to which I refer is that to which the body would be subject if its motion were continued along the plane AB extended, while the retardation is that which the body would encounter if its motion were deflected along the plane BC inclined upwards; but, upon the horizontal plane GH, the body would maintain a uniform velocity equal to that which it had acquired at B after fall from A; moreover this velocity is such that, during an interval of time equal to the time of descent through AB, the body will traverse a horizontal distance equal to twice AB. Now let us imagine this same body to move with the same uniform speed along the plane BC so that here also during a time-interval equal to that of descent along AB, it will traverse along BC extended a distance twice AB; but let us suppose that, at the very instant the body begins its ascent it is subjected, by its very nature, to the same influences which

surrounded

surrounded it during its descent from A along AB, namely, it descends from rest under the same acceleration as that which was effective in AB, and it traverses, during an equal interval of time, the same distance along this second plane as it did along AB; it is clear that, by thus superposing upon the body a uniform motion of ascent and an accelerated motion of descent, it will be carried along the plane BC as far as the point C where these two velocities become equal.

If now we assume any two points D and E, equally distant from the vertex B, we may then infer that the descent along BD takes place in the same time as the ascent along BE. Draw DF parallel to BC; we know that, after descent along AD, the body will ascend along DF; or, if, on reaching D, the body is carried along the horizontal DE, it will reach E with the same momentum [*impetus*] with which it left D; hence from E the body will ascend as far as C, proving that the velocity at E is the same as that at D.

From this we may logically infer that a body which descends
[245]
along any inclined plane and continues its motion along a plane inclined upwards will, on account of the momentum acquired, ascend to an equal height above the horizontal; so that if the descent is along AB the body will be carried up the plane BC as far as the horizontal line ACD: and this is true whether the inclinations of the

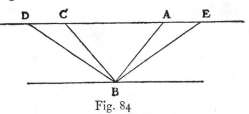

Fig. 84

planes are the same or different, as in the case of the planes AB and BD. But by a previous postulate [p. 184] the speeds acquired by fall along variously inclined planes having the same vertical height are the same. If therefore the planes EB and BD have the same slope, the descent along EB will be able to drive the body along BD as far as D; and since this propulsion comes from the speed acquired on reaching
the

the point B, it follows that this speed at B is the same whether the body has made its descent along AB or EB. Evidently then the body will be carried up BD whether the descent has been made along AB or along EB. The time of ascent along BD is however greater than that along BC, just as the descent along EB occupies more time than that along AB; moreover it has been demonstrated that the ratio between the lengths of these times is the same as that between the lengths of the planes. We must next discover what ratio exists between the distances traversed in equal times along planes of different slope, but of the same elevation, that is, along planes which are included between the same parallel horizontal lines. This is done as follows:

THEOREM XV, PROPOSITION XXIV

Given two parallel horizontal planes and a vertical line connecting them; given also an inclined plane passing through the lower extremity of this vertical line; then, if a body fall freely along the vertical line and have its motion reflected along the inclined plane, the distance which it will traverse along this plane, during a time equal to that of the vertical fall, is greater than once but less than twice the vertical line.

Let BC and HG be the two horizontal planes, connected by the perpendicular AE; also let EB represent the inclined plane

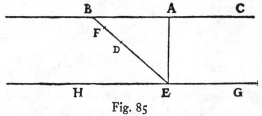

Fig. 85

along which the motion takes place after the body has fallen along AE and has been reflected from E towards B. Then, I say, that, during a time equal to that of fall along AE, the body will ascend the inclined plane through a distance which is
greater

greater than AE but less than twice AE. Lay off ED equal to AE and choose F so that EB:BD = BD:BF. First we shall [245] show that F is the point to which the moving body will be carried after reflection from E towards B during a time equal to that of fall along AE; and next we shall show that the distance EF is greater than EA but less than twice that quantity.

Let us agree to represent the time of fall along AE by the length AE, then the time of descent along BE, or what is the same thing, ascent along EB will be represented by the distance EB.

Now, since DB is a mean proportional between EB and BF, and since BE is the time of descent for the entire distance BE, it follows that BD will be the time of descent through BF, while the remainder DE will be the time of descent along the remainder FE. But the time of descent along the fall from rest at B is the same as the time of ascent from E to F after reflection from E with the speed acquired during fall either through AE or BE. Therefore DE represents the time occupied by the body in passing from E to F, after fall from A to E and after reflection along EB. But by construction ED is equal to AE. This concludes the first part of our demonstration.

Now since the whole of EB is to the whole of BD as the portion DB is to the portion BF, we have the whole of EB is to the whole of BD as the remainder ED is to the remainder DF; but EB>BD and hence ED>DF, and EF is less than twice DE or AE.   Q. E. D.

The same is true when the initial motion occurs, not along a perpendicular, but upon an inclined plane: the proof is also the same provided the upward sloping plane is less steep, i. e., longer, than the downward sloping plane.

THEOREM XVI, PROPOSITION XXV

If descent along any inclined plane is followed by motion along a horizontal plane, the time of descent along the inclined plane bears to the time required to traverse any assigned length of the horizontal plane the same ratio which

twice

twice the length of the inclined plane bears to the given horizontal length.

Let CB be any horizontal line and AB an inclined plane; after descent along AB let the motion continue through the assigned horizontal distance BD. Then, I say, the time of descent along AB bears to the time spent in traversing BD the same ratio which twice AB bears to BD. For, lay off BC equal to twice AB then it follows, from a previous proposition, that the time of descent along AB is equal to the time required to traverse BC; but the time along BC is to the time along DB as the length CB is to the length BD. Hence the time of descent along AB

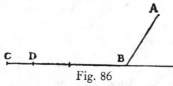

Fig. 86

[247]

is to the time along BD as twice the distance AB is to the distance BD.

Q. E. D.

Problem X, Proposition XXVI

Given a vertical height joining two horizontal parallel lines; given also a distance greater than once and less than twice this vertical height, it is required to pass through the foot of the given perpendicular an inclined plane such that, after fall through the given vertical height, a body whose motion is deflected along the plane will traverse the assigned distance in a time equal to the time of vertical fall.

Let AB be the vertical distance separating two parallel horizontal lines AO and BC; also let FE be greater than once and less than twice BA. The problem is to pass a plane through B, extending to the upper horizontal line, and such that a body, after having fallen from A to B, will, if its motion be deflected along the inclined plane, traverse a distance equal to EF in a time equal to that of fall along AB. Lay off ED equal to AB; then the remainder DF will be less than AB since the entire length EF is less than twice this quantity; also lay off DI equal to DF, and choose the point X such that EI:ID = DF:FX; from B, draw the plane BO equal in length to EX. Then, I say, that

that the plane BO is the one along which, after fall through AB, a body will traverse the assigned distance FE in a time equal to the time of fall through AB. Lay off BR and RS equal to ED and DF respectively; then since EI:ID =DF:FX, we have, *componendo*, ED:DI =DX:XF =ED:DF =EX:XD =BO:OR =

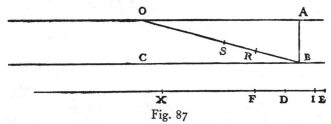

Fig. 87

RO:OS. If we represent the time of fall along AB by the length AB, then OB will represent the time of descent along

[248]

OB, and RO will stand for the time along OS, while the remainder BR will represent the time required for a body starting from rest at O to traverse the remaining distance SB. But the time of descent along SB starting from rest at O is equal to the time of ascent from B to S after fall through AB. Hence BO is that plane, passing through B, along which a body, after fall through AB, will traverse the distance BS, equal to the assigned distance EF, in the time-interval BR or BA.   Q. E. F.

THEOREM XVII, PROPOSITION XXVII

If a body descends along two inclined planes of different lengths but of the same vertical height, the distance which it will traverse, in the lower part of the longer plane, during a time-interval equal to that of descent over the shorter plane, is equal to the length of the shorter plane plus a portion of it to which the shorter plane bears the same ratio which the longer plane bears to the excess of the longer over the shorter plane.

Let AC be the longer plane, AB, the shorter, and AD the common elevation; on the lower part of AC lay off CE equal to

to

to AB. Choose F such that CA:AE = CA:CA–AB = CE:EF. Then, I say, that FC is that distance which will, after fall from A, be traversed during a time-interval equal to that required for

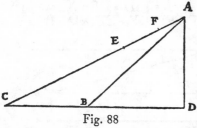

descent along AB. For since CA:AE = CE:EF, it follows that the remainder EA: the remainder AF = CA : AE. Therefore AE is a mean proportional between AC and AF. Accordingly if the length AB is employed to measure the time of fall along AB, then

Fig. 88

the distance AC will measure the time of descent through AC; but the time of descent through AF is measured by the length AE, and that through FC by EC. Now EC = AB; and hence follows the proposition.

[249]
Problem XI, Proposition XXVIII

Let AG be any horizontal line touching a circle; let AB be the diameter passing through the point of contact; and let AE and EB represent any two chords. The problem is to determine what ratio the time of fall through AB bears to the time of descent over both AE and EB. Extend BE till it meets the tangent at G, and draw AF so as to bisect the angle BAE. Then, I say, the time through AB is to the sum of the times along AE and EB as the length AE is to the sum of the lengths AE and EF. For since the angle FAB is equal to the angle FAE, while the angle EAG is equal to the angle ABF it

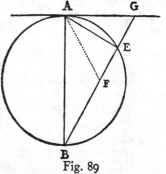

Fig. 89

follows that the entire angle GAF is equal to the sum of the angles FAB and ABF. But the angle GFA is also equal to the sum of these two angles. Hence the length GF is equal to the length GA

GA; and since the rectangle BG.GE is equal to the square of GA, it will also be equal to the square of GF, or BG:GF = GF:GE.   If now we agree to represent the time of descent along AE by the length AE, then the length GE will represent the time of descent along GE, while GF will stand for the time of descent through the entire distance GB; so also EF will denote the time through EB after fall from G or from A along AE.   Consequently the time along AE, or AB, is to the time along AE and EB as the length AE is to AE+EF.   Q. E. D.

A shorter method is to lay off GF equal to GA, thus making GF a mean proportional between BG and GE.   The rest of the proof is as above.

THEOREM XVIII, PROPOSITION XXIX

Given a limited horizontal line, at one end of which is erected a limited vertical line whose length is equal to one-half the given horizontal line; then a body, falling through this given height and having its motion deflected into a horizontal direction, will traverse the given horizontal distance and vertical line in less time than it will any other vertical distance plus the given horizontal distance.

[250]

Let BC be the given distance in a horizontal plane; at the end B erect a perpendicular, on which lay off BA equal to half

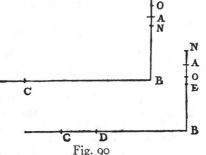

Fig. 90

BC.   Then, I say, that the time required for a body, starting from rest at A, to traverse the two distances, AB and BC, is the least of all possible times in which this same distance BC together with a vertical portion, whether greater or less than AB, can be traversed.

Lay off EB greater than AB, as in the first figure, and less than

than AB, as in the second. It must be shown that the time required to traverse the distance EB plus BC is greater than that required for AB plus BC. Let us agree that the length AB shall represent the time along AB, then the time occupied in traversing the horizontal portion BC will also be AB, seeing that BC = 2AB; consequently the time required for both AB and BC will be twice AB. Choose the point O such that EB: BO = BO:BA, then BO will represent the time of fall through EB. Again lay off the horizontal distance BD equal to twice BE; whence it is clear that BO represents the time along BD after fall through EB. Select a point N such that DB:BC = EB:BA = OB:BN. Now since the horizontal motion is uniform and since OB is the time occupied in traversing BD, after fall from E, it follows that NB will be the time along BC after fall through the same height EB. Hence it is clear that OB plus BN represents the time of traversing EB plus BC; and, since twice BA is the time along AB plus BC, it remains to be shown that OB+BN>2BA.

But since EB:BO = BO:BA, it follows that EB:BA = $\overline{OB}^2$: $\overline{BA}^2$. Moreover since EB:BA = OB:BN it follows that OB:BN = $\overline{OB}^2$:$\overline{BA}^2$. But OB:BN = (OB:BA)(BA:BN), and therefore AB:BN = OB:BA, that is, BA is a mean proportional between BO and BN. Consequently OB+BN>2BA.   Q. E. D.

[251]
THEOREM XIX, PROPOSITION XXX

A perpendicular is let fall from any point in a horizontal line; it is required to pass through any other point in this same horizontal line a plane which shall cut the perpendicular and along which a body will descend to the perpendicular in the shortest possible time. Such a plane will cut from the perpendicular a portion equal to the distance of the assumed point in the horizontal from the upper end of the perpendicular.

Let AC be any horizontal line and B any point in it from which is dropped the vertical line BD. Choose any point C in the horizontal line and lay off, on the vertical, the distance BE equal

equal to BC; join C and E. Then, I say, that of all inclined planes that can be passed through C, cutting the perpendicular, CE is that one along which the descent to the perpendicular is accomplished in the shortest time. For, draw the plane CF cutting the vertical above E, and the plane CG cutting the vertical below E; and draw IK, a parallel vertical line, touching at C a circle described with BC as radius. Let EK be drawn parallel to CF, and extended to meet the tangent, after cutting the circle at L. Now it is clear that the time of fall along LE is equal to the time along CE; but the time along KE is greater than along LE; therefore the time along KE is greater than along CE. But the time along KE is equal to the time along CF, since they have the same length and the same slope; and, in like manner, it follows that the planes CG and IE, having the same length and the same slope, will be traversed in equal times.

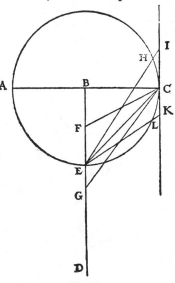

Fig. 91

Also, since HE < IE, the time along HE will be less than the time along IE. Therefore also the time along CE (equal to the time along HE), will be shorter than the time along IE.   Q. E. D.

Theorem XX, Proposition XXXI

If a straight line is inclined at any angle to the horizontal and if, from any assigned point in the horizontal, a plane of quickest descent is to be drawn to the inclined line, that plane will be the one which bisects the angle contained

[252]

between two lines drawn from the given point, one perpendicular

pendicular to the horizontal line, the other perpendicular to the inclined line.

Let CD be a line inclined at any angle to the horizontal AB; and from any assigned point A in the horizontal draw AC perpendicular to AB, and AE perpendicular to CD; draw FA so as to bisect the angle CAE. Then, I say, that of all the planes which can be drawn through the point A, cutting the line CD

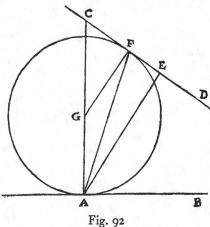

Fig. 92

at any points whatsoever AF is the one of quickest descent [*in quo tempore omnium brevissimo fiat descensus*]. Draw FG parallel to AE; the alternate angles GFA and FAE will be equal; also the angle EAF is equal to the angle FAG. Therefore the sides GF and GA of the triangle FGA are equal. Accordingly if we describe a circle about G as center, with GA as radius, this circle will pass through the point F, and will touch the horizontal at the point A and the inclined line at F; for GFC is a right angle, since GF and AE are parallel. It is clear therefore that all lines drawn from A to the inclined line, with the single exception of FA, will extend beyond the circumference of the circle, thus requiring more time to traverse any of them than is needed for FA.   Q. E. D.

LEMMA

If two circles one lying within the other are in contact, and if any straight line be drawn tangent to the inner circle, cutting the outer circle, and if three lines be drawn from the point at which the circles are in contact to three points on the tangential straight line, namely, the point of tangency on the inner circle and the two points where the straight

straight line extended cuts the outer circle, then these three
lines will contain equal angles at the point of contact.

Let the two circles touch each other at the point A, the center
of the smaller being at B, the center of the larger at C. Draw

[253]

the straight line FG touching the inner circle at H, and cutting
the outer at the points F and G; also draw the three lines AF,
AH, and AG. Then, I say, the angles contained by these lines,
FAH and GAH, are equal. Pro-
long AH to the circumference at
I; from the centers of the circles,
draw BH and CI; join the centers
B and C and extend the line until
it reaches the point of contact at
A and cuts the circles at the
points O and N. But now the
lines BH and CI are parallel, because the angles ICN and HBO
are equal, each being twice the
angle IAN. And since BH, drawn
from the center to the point of

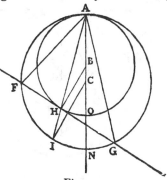

Fig. 93

contact is perpendicular to FG, it follows that CI will also be
perpendicular to FG and that the arc FI is equal to the arc IG;
consequently the angle FAI is equal to the angle IAG. Q. E. D.

Theorem XXI, Proposition XXXII

If in a horizontal line any two points are chosen and if
through one of these points a line be drawn inclined towards
the other, and if from this other point a straight line is
drawn to the inclined line in such a direction that it cuts
off from the inclined line a portion equal to the distance
between the two chosen points on the horizontal line, then
the time of descent along the line so drawn is less than along
any other straight line drawn from the same point to the
same inclined line. Along other lines which make equal
angles on opposite sides of this line, the times of descent are
the same.

Let

Let A and B be any two points on a horizontal line: through B draw an inclined straight line BC, and from B lay off a distance BD equal to BA; join the points A and D. Then, I say, the time of descent along AD is less than along any other line drawn from A to the inclined line BC. From the point A draw AE perpendicular to BA; and from the point D draw DE perpendicular to BD, intersecting AE at E. Since in the isosceles triangle ABD, we have the angles BAD and BDA equal,

[254]

their complements DAE and EDA are equal. Hence if, with E as center and EA as radius, we describe a circle it will pass through D and will touch the lines BA and BD at the points A and D. Now since A is the end of the vertical line AE, the descent along AD will occupy less time than along any other line drawn from the extremity A to the line BC and extending beyond the circumference of the circle; which concludes the first part of the proposition.

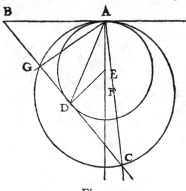

Fig. 94

If however, we prolong the perpendicular line AE, and choose any point F upon it, about which as center, we describe a circle of radius FA, this circle, AGC, will cut the tangent line in the points G and C. Draw the lines AG and AC which will according to the preceding lemma, deviate by equal angles from the median line AD. The time of descent along either of these lines is the same, since they start from the highest point A, and terminate on the circumference of the circle AGC.

PROBLEM XII, PROPOSITION XXXIII

Given a limited vertical line and an inclined plane of equal height, having a common upper terminal; it is required to find a point on the vertical line, extended upwards, from which

which a body will fall and, when deflected along the inclined plane, will traverse it in the same time-interval which is required for fall, from rest, through the given vertical height.

Let AB be the given limited vertical line and AC an inclined plane having the same altitude. It is required to find on the vertical BA, extended above A, a point from which a falling body will traverse the distance AC in the same time which is spent in falling, from rest at A, through the given vertical line AB. Draw the line DCE at right angles to AC, and lay off CD equal to AB; also join the points A and D; then the angle ADC will be greater than the angle CAD, since the side CA is greater than either AB or CD. Make the angle DAE equal to the angle

[255]

ADE, and draw EF perpendicular to AE; then EF will cut the inclined plane, ex-

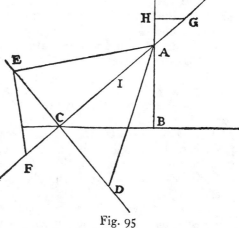

tended both ways, at F. Lay off AI and AG each equal to CF; through G draw the horizontal line GH. Then, I say, H is the point sought.

For, if we agree to let the length AB represent the time of fall along the verti-cal AB, then AC will likewise represent the time of descent from rest at A, along AC;

Fig. 95

and since, in the right-angled triangle AEF, the line EC has been drawn from the right angle at E perpendicular to the base AF, it follows that AE will be a mean proportional between FA and AC, while CE will be a mean proportional between AC and CF, that is between CA and AI. Now, since AC represents the time of descent from A along AC, it follows that AE will be the time along the entire distance AF, and EC the time along AI. But since

since in the isosceles triangle AED the side EA is equal to the side ED it follows that ED will represent the time of fall along AF, while EC is the time of fall along AI. Therefore CD, that is AB, will represent the time of fall, from rest at A, along IF; which is the same as saying that AB is the time of fall, from G or from H, along AC.   E. F.

PROBLEM XIII, PROPOSITION XXXIV

Given a limited inclined plane and a vertical line having their highest point in common, it is required to find a point in the vertical line extended such that a body will fall from it and then traverse the inclined plane in the same time which is required to traverse the inclined plane alone starting from rest at the top of said plane.

Let AC and AB be an inclined plane and a vertical line respectively, having a common highest point at A. It is required to find a point in the vertical line, above A, such that a body, falling from it and afterwards having its motion directed along AB, will traverse both the assigned part of the vertical

[256]

line and the plane AB in the same time which is required for the plane AB alone, starting from rest at A. Draw BC a horizontal line and lay off AN equal to AC; choose the point L so that AB:BN =AL:LC, and lay off AI equal to AL; choose the point E such that CE, laid off on the vertical AC produced, will be a third proportional to AC and BI. Then, I say, CE is the distance sought; so that, if the vertical line is extended above A and if a portion AX is laid off equal to CE, then a body falling from X will traverse both the distances, XA and AB, in the same time as that required, when starting from A, to traverse AB alone.

Draw XR parallel to BC and intersecting BA produced in R; next draw ED parallel to BC and meeting BA produced in D; on AD as diameter describe a semicircle; from B draw BF perpendicular to AD, and prolong it till it meets the circumference of the circle; evidently FB is a mean proportional between AB and BD, while FA is a mean proportional between
DA

DA and AB. Take BS equal to BI and FH equal to FB. Now since AB:BD =AC:CE and since BF is a mean proportional

[257]

between AB and BD, while BI is a mean proportional between AC and CE, it follows that BA:AC =FB:BS, and since BA: AC =BA:BN =FB:BS we shall have, *convertendo*, BF:FS = AB:BN =AL:LC. Consequently the rectangle formed by FB

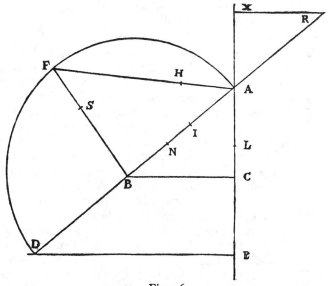

Fig. 96

and CL is equal to the rectangle whose sides are AL and SF; moreover, this rectangle AL.SF is the excess of the rectangle AL.FB, or AI.BF, over the rectangle AI.BS, or AI.IB. But the rectangle FB.LC is the excess of the rectangle AC.BF over the rectangle AL.BF; and moreover the rectangle AC.BF is equal to the rectangle AB.BI since BA:AC =FB:BI; hence the excess of the rectangle AB.BI over the rectangle AI.BF, or AI.FH, is equal to the excess of the rectangle AI.FH over the rectangle AI.IB; therefore twice the rectangle AI.FH is equal to the sum

of

of the rectangles AB.BI and AI.IB, or $2AI.FH = 2AI.IB + \overline{BI}^2$. Add $\overline{AI}^2$ to each side, then $2AI.IB + \overline{BI}^2 + \overline{AI}^2 = \overline{AB}^2 = 2AI.FH + AI^2$. Again add $\overline{BF}^2$ to each side, then $AB^2 + BF^2 = \overline{AF}^2 = 2AI.FH + \overline{AI}^2 + \overline{BF}^2 = 2AI.FH + \overline{AI}^2 + \overline{FH}^2$. But $\overline{AF}^2 = 2AH.HF + \overline{AH}^2 + \overline{HF}^2$; and hence $2AI.FH + \overline{AI}^2 + \overline{FH}^2 = 2AH.HF + \overline{AH}^2 + \overline{HF}^2$. Subtracting $\overline{HF}^2$ from each side we have $2AI.FH + \overline{AI}^2 = 2AH.HF + \overline{AH}^2$. Since now FH is a factor common to both rectangles, it follows that AH is equal to AI; for if AH were either greater or smaller than AI, then the two rectangles AH.HF plus the square of HA would be either larger or smaller than the two rectangles AI.FH plus the square of IA, a result which is contrary to what we have just demonstrated.

If now we agree to represent the time of descent along AB by the length AB, then the time through AC will likewise be measured by AC; and IB, which is a mean proportional between AC and CE, will represent the time through CE, or XA, from rest at X. Now, since AF is a mean proportional between DA and AB, or between RB and AB, and since BF, which is equal to FH, is a mean proportional between AB and BD, that is between AB and AR, it follows, from a preceding proposition [Proposition XIX, corollary], that the difference AH represents the time of descent along AB either from rest at R or after fall from X, while the time of descent along AB, from rest at A, is measured by the length AB. But as has just been shown, the time of fall through XA is measured by IB, while the time of descent along AB, after fall, through RA or through XA, is IA. Therefore the time of descent through XA plus AB is measured by the length AB, which, of course, also measures the time of descent, from rest at A, along AB alone.   Q. E. F.

[258]
PROBLEM XIV, PROPOSITION XXXV

Given an inclined plane and a limited vertical line, it is required to find a distance on the inclined plane which a body, starting from rest, will traverse in the same time as that needed to traverse both the vertical and the inclined plane.
Let

Let AB be the vertical line and BC the inclined plane. It is required to lay off on BC a distance which a body, starting from rest, will traverse in a time equal to that which is occupied by fall through the vertical AB and by descent of the plane. Draw the horizontal line AD, which intersects at E the prolongation of the inclined plane CB; lay off BF equal to BA, and about E as center, with EF as radius describe the circle FIG. Prolong FE until it intersects the circumference at G. Choose a point H such that GB:BF = BH:HF. Draw the line HI tangent to the

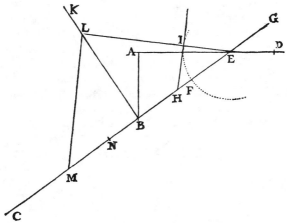

Fig. 97

circle at I. At B draw the line BK perpendicular to FC, cutting the line EIL at L; also draw LM perpendicular to EL and cutting BC at M. Then, I say, BM is the distance which a body, starting from rest at B, will traverse in the same time which is required to descend from rest at A through both distances, AB and BM. Lay off EN equal to EL; then since GB:BF = BH:HF, we shall have, *permutando*, GB:BH = BF:HF, and, *dividendo*, GH:BH = BH:HF. Consequently the rectangle GH.HF is equal to the square on BH; but this same rectangle is also equal to the square on HI; therefore BH is equal to HI. Since, in the quadrilateral ILBH, the sides HB and HI are
equal

equal and since the angles at B and I are right angles, it follows that the sides BL and LI are also equal: but EI = EF; therefore

[259]

the total length LE, or NE, is equal to the sum of LB and EF. If we subtract the common part EF, the remainder FN will be equal to LB: but, by construction, FB = BA and, therefore, LB = AB + BN. If again we agree to represent the time of fall through AB by the length AB, then the time of descent along EB will be measured by EB; moreover since EN is a mean proportional between ME and EB it will represent the time of descent along the whole distance EM; therefore the difference of these distances, BM, will be traversed, after fall from EB, or AB, in a time which is represented by BN. But having already assumed the distance AB as a measure of the time of fall through AB, the time of descent along AB and BM is measured by AB + BN. Since EB measures the time of fall, from rest at E, along EB, the time from rest at B along BM will be the mean proportional between BE and BM, namely, BL. The time there-

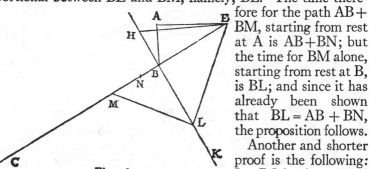

Fig. 98

fore for the path AB + BM, starting from rest at A is AB + BN; but the time for BM alone, starting from rest at B, is BL; and since it has already been shown that BL = AB + BN, the proposition follows.

Another and shorter proof is the following: Let BC be the inclined plane and BA the vertical; at B draw a perpendicular to EC, extending it both ways; lay off BH equal to the excess of BE over BA; make the angle HEL equal to the angle BHE; prolong EL until it cuts BK in L; at L draw LM perpendicular to EL and extend it till it meets BC in M; then, I say, BM is the portion of BC sought. For, since the angle MLE is a right angle, BL will be a mean proportional between MB and BE, while

while LE is a mean proportional between ME and BE; lay off EN equal to LE; then NE = EL = LH, and HB = NE−BL. But also HB = NE−(NB+BA); therefore BN+BA = BL. If now we assume the length EB as a measure of the time of descent along EB, the time of descent, from rest at B, along BM will be represented by BL; but, if the descent along BM is from rest at E or at A, then the time of descent will be measured by BN; and AB will measure the time along AB. Therefore the time required to traverse AB and BM, namely, the sum of the distances AB and BN, is equal to the time of descent, from rest at B, along BM alone.   Q. E. F.

[260]
LEMMA

Let DC be drawn perpendicular to the diameter BA; from the extremity B draw the line BED at random; draw the line FB. Then, I say, FB is a mean proportional between DB and BE. Join the points E and F. Through B, draw the tangent BG which will be parallel to CD. Now, since the angle DBG is equal to the angle FDB, and since the alternate angle of GBD is equal to EFB, it follows that the triangles FDB and FEB are similar and hence BD:BF = FB:BE.

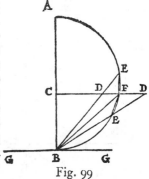

Fig. 99

LEMMA

Let AC be a line which is longer than DF, and let the ratio of AB to BC be greater than that of DE to EF. Then, I say, AB is greater than DE. For, if AB bears to BC a ratio greater than that of DE to EF, then DE will bear to some length shorter than EF, the same ratio which AB bears to BC. Call this length EG; then since AB:BC = DE:EG, it follows, *componendo et convertendo,*

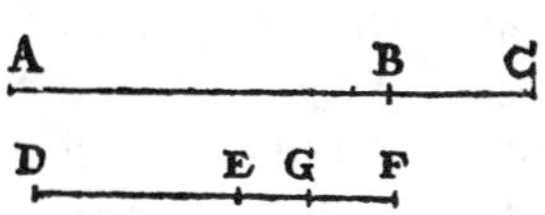

Fig. 100

vertendo, that CA:AB =GD:DE. But since CA is greater than GD, it follows that BA is greater than DE.

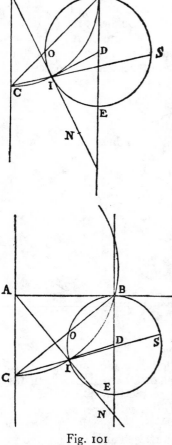

Fig. 101

LEMMA

Let ACIB be the quadrant of a circle; from B draw BE parallel to AC; about any point in the line BE describe a circle BOES, touching AB at B and intersecting the circumference of the quadrant at I. Join the points C and B; draw the line CI, prolonging it to S. Then, I say, the line CI is always less than CO. Draw the line AI touching the circle BOE. Then,

[261]

if the line DI be drawn, it will be equal to DB; but, since DB touches the quadrant, DI will also be tangent to it and will be at right angles to AI; thus AI touches the circle BOE at I. And since the angle AIC is greater than the angle ABC, subtending as it does a larger arc, it follows that the angle SIN is also greater than the angle ABC. Wherefore the arc IES is greater than the arc BO, and the line CS, being nearer the center, is longer than CB. Consequently CO is greater than CI, since SC: CB =OC:CI.

This result would be all the more marked if, as in the second figure, the arc BIC were less than a quadrant. For the perpendicular DB would then cut the circle CIB; and so also would DI

DI which is equal to BD; the angle DIA would be obtuse and therefore the line AIN would cut the circle BIE. Since the angle ABC is less than the angle AIC, which is equal to SIN, and still less than the angle which the tangent at I would make with the line Sl, it follows that the arc SEI is far greater than the arc BO; whence, etc.   Q. E. D.

THEOREM XXII, PROPOSITION XXXVI

If from the lowest point of a vertical circle, a chord is drawn subtending an arc not greater than a quadrant, and if from the two ends of this chord two other chords be drawn to any point on the arc, the time of descent along the two latter chords will be shorter than along the first, and shorter also, by the same amount, than along the lower of these two latter chords.

[262]

Let CBD be an arc, not exceeding a quadrant, taken from a vertical circle whose lowest point is C; let CD be the chord [*planum elevatum*] sub-tending this arc, and let there be two other chords drawn from C and D to any point B on the arc. Then, I say, the time of descent along the two chords [*plana*] DB and BC is shorter than along DC alone, or along BC alone, starting from rest at B. Through the point D, draw the horizontal line MDA cutting CB extended at

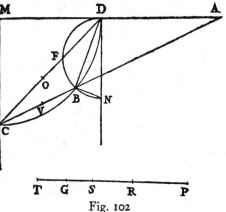

Fig. 102

A: draw DN and MC at right angles to MD, and BN at right angles to BD; about the right-angled triangle DBN describe the semicircle DFBN, cutting DC at F. Choose the point O such that DO will be a mean proportional between CD and DF; in like manner

manner select V so that AV is a mean proportional between CA and AB. Let the length PS represent the time of descent along the whole distance DC or BC, both of which require the same time. Lay off PR such that CD:DO =*time*PS. *time*PR. Then PR will represent the time in which a body, starting from D, will traverse the distance DF, while RS will measure the time in which the remaining distance, FC, will be traversed. But since PS is also the time of descent, from rest at B, along BC, and if we choose T such that BC:CD =PS:PT then PT will measure the time of descent from A to C, for we have already shown [Lemma] that DC is a mean proportional between AC and CB. Finally choose the point G such that CA:AV =PT:PG, then PG will be the time of descent from A to B, while GT will be the residual time of descent along BC following descent from A to B. But, since the diameter, DN, of the circle DFN is a vertical line, the chords DF and DB will be traversed in equal times; wherefore if one can prove that a body will traverse BC, after descent along DB, in a shorter time than it will FC after descent along DF he will have proved the theorem. But a body descending from D along DB will traverse BC in the same time as if it had come from A along AB, seeing that the body acquires the same

[263]

momentum in descending along DB as along AB. Hence it remains only to show that descent along BC after AB is quicker than along FC after DF. But we have already shown that GT represents the time along BC after AB; also that RS measures the time along FC after DF. Accordingly it must be shown that RS is greater than GT, which may be done as follows: Since SP:PR =CD:DO, it follows, *invertendo et convertendo*, that RS:SP =OC:CD; also we have SP:PT =DC:CA. And since TP:PG =CA:AV, it follows, *invertendo*, that PT:TG = AC:CV, therefore, *ex æquali*, RS:GT =OC:CV. But, as we shall presently show, OC is greater than CV; hence the time RS is greater than the time GT, which was to be shown. Now, since [Lemma] CF is greater than CB and FD smaller than BA, it follows that CD:DF>CA:AB. But CD:DF =CO:OF, seeing that CD:DO =DO:DF; and CA:AB =$\overline{CV}^2$:$\overline{VB}^2$. Therefore

fore CO:OF>CV:VB, and, according to the preceding lemma, CO>CV. Besides this it is clear that the time of descent along DC is to the time along DBC as DOC is to the sum of DO and CV.

SCHOLIUM

From the preceding it is possible to infer that the path of quickest descent [*lationem omnium velocissimam*] from one point to another is not the shortest path, namely, a straight line, but the arc of a circle.* In the quadrant BAEC, having the side BC vertical, divide the arc AC into any number of equal parts, AD, DE, EF, FG, GC, and from C draw straight lines to the points A, D, E, F, G; draw also the straight lines AD, DE, EF, FG, GC. Evidently descent along the path ADC is quicker [264] than along AC alone or along DC from rest at D. But a body, starting from rest at A, will traverse DC more quickly than the path ADC; while, if it starts from rest at A, it will traverse the path DEC in a shorter time than DC alone. Hence descent along the three

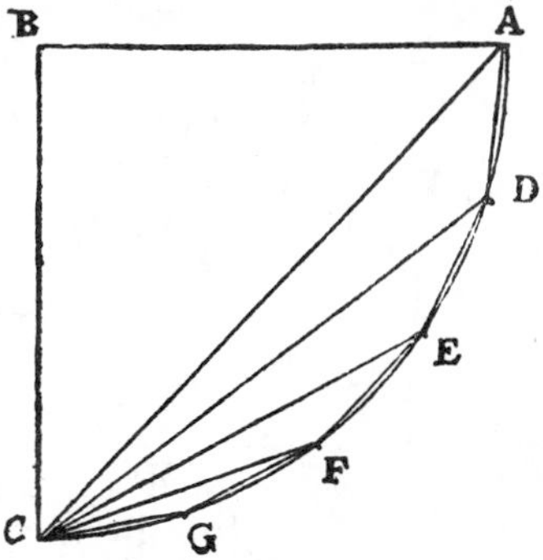

Fig. 103

chords, ADEC, will take less time than along the two chords ADC. Similarly, following descent along ADE, the time required to traverse EFC is less than that needed for EC alone. Therefore descent is more rapid along the four chords ADEFC than along the three ADEC. And finally a body, after descent along ADEF, will traverse the two chords, FGC, more quickly than FC alone. Therefore, along the five chords, ADEFGC, descent will be more rapid than along the four, ADEFC. Consequently

---

* It is well known that the first correct solution for the problem of quickest descent, under the condition of a constant force was given by John Bernoulli (1667–1748). [*Trans.*]

the nearer the inscribed polygon approaches a circle the shorter is the time required for descent from A to C.

What has been proven for the quadrant holds true also for smaller arcs; the reasoning is the same.

PROBLEM XV, PROPOSITION XXXVII

Given a limited vertical line and an inclined plane of equal altitude; it is required to find a distance on the inclined plane which is equal to the vertical line and which is traversed in an interval equal to the time of fall along the vertical line.

Let AB be the vertical line and AC the inclined plane. We must locate, on the inclined plane, a distance equal to the vertical

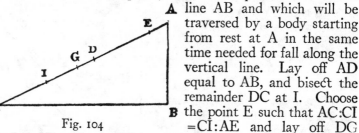

line AB and which will be traversed by a body starting from rest at A in the same time needed for fall along the vertical line. Lay off AD equal to AB, and bisect the remainder DC at I. Choose the point E such that AC:CI =CI:AE and lay off DG

Fig. 104

equal to AE. Clearly EG is equal to AD, and also to AB. And further, I say that EG is that distance which will be traversed by a body, starting from rest at A, in the same time which is required for that body to fall through the distance AB. For since AC:CI =CI:AE =ID:DG, we have, *convertendo*, CA: AI =DI:IG. And since the whole of CA is to the whole of AI as the portion CI is to the portion IG, it follows that the re-

[265]

mainder IA is to the remainder AG as the whole of CA is to the whole of AI. Thus AI is seen to be a mean proportional between CA and AG, while CI is a mean proportional between CA and AE. If therefore the time of fall along AB is represented by the length AB, the time along AC will be represented by AC, while CI, or ID, will measure the time along AE. Since AI is a mean proportional between CA and AG, and since CA is a

measure

measure of the time along the entire distance AC, it follows that AI is the time along AG, and the difference IC is the time along the difference GC; but DI was the time along AE. Consequently the lengths DI and IC measure the times along AE and CG respectively. Therefore the remainder DA represents the time along EG, which of course is equal to the time along AB.

Q. E. F.

COROLLARY

From this it is clear that the distance sought is bounded at each end by portions of the inclined plane which are traversed in equal times.

PROBLEM XVI, PROPOSITION XXXVIII

Given two horizontal planes cut by a vertical line, it is required to find a point on the upper part of the vertical line from which bodies may fall to the horizontal planes and there, having their motion deflected into a horizontal direction, will, during an interval equal to the time of fall, traverse distances which bear to each other any assigned ratio of a smaller quantity to a larger.

Let CD and BE be the horizontal planes cut by the vertical ACB, and let the ratio of the smaller quantity to the larger be that of N to FG. It is required to find in the upper part of the vertical line, AB, a point from which a body falling to the plane CD and there having its motion deflected along this plane, will traverse, during an interval equal to its time of fall a distance such that if another body, falling from this same point to the plane BE, there have its motion deflected along this plane and continued during an interval equal to its time of fall, will traverse a distance which bears to the former distance the

[266]

ratio of FG to N. Lay off GH equal to N, and select the point L so that FH:HG = BC:CL. Then, I say, L is the point sought. For, if we lay off CM equal to twice CL, and draw the line LM cutting the plane BE at O, then BO will be equal to twice

BL

BL. And since FH:HG = BC:CL, we have, *componendo et convertendo*, HG:GF = N:GF = CL:LB = CM:BO. It is clear that, since CM is double the distance LC, the space CM is that which a body falling from L through LC will traverse in the plane CD; and, for the same reason, since BO is twice the distance BL, it is clear that BO is the distance which a body,

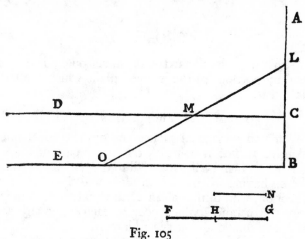

Fig. 105

after fall through LB, will traverse during an interval equal to the time of its fall through LB.   Q. E. F.

SAGR. Indeed, I think we may concede to our Academician, without flattery, his claim that in the principle [*principio*, i. e., accelerated motion] laid down in this treatise he has established a new science dealing with a very old subject. Observing with what ease and clearness he deduces from a single principle the proofs of so many theorems, I wonder not a little how such a question escaped the attention of Archimedes, Apollonius, Euclid and so many other mathematicians and illustrious philosophers, especially since so many ponderous tomes have been devoted to the subject of motion.

[267]

SALV. There is a fragment of Euclid which treats of motion, but

but in it there is no indication that he ever began to investigate the property of acceleration and the manner in which it varies with slope. So that we may say the door is now opened, for the first time, to a new method fraught with numerous and wonderful results which in future years will command the attention of other minds.

SAGR. I really believe that just as, for instance, the few properties of the circle proven by Euclid in the Third Book of his Elements lead to many others more recondite, so the principles which are set forth in this little treatise will, when taken up by speculative minds, lead to many another more remarkable result; and it is to be believed that it will be so on account of the nobility of the subject, which is superior to any other in nature.

During this long and laborious day, I have enjoyed these simple theorems more than their proofs, many of which, for their complete comprehension, would require more than an hour each; this study, if you will be good enough to leave the book in my hands, is one which I mean to take up at my leisure after we have read the remaining portion which deals with the motion of projectiles; and this if agreeable to you we shall take up to-morrow.

SALV. I shall not fail to be with you.

END OF THE THIRD DAY.

[268]

FOURTH DAY

ALVIATI. Once more, Simplicio is here on time; so let us without delay take up the question of motion. The text of our Author is as follows:

THE MOTION OF PROJECTILES

In the preceding pages we have discussed the properties of uniform motion and of motion naturally accelerated along planes of all inclinations. I now propose to set forth those properties which belong to a body whose motion is compounded of two other motions, namely, one uniform and one naturally accelerated; these properties, well worth knowing, I propose to demonstrate in a rigid manner. This is the kind of motion seen in a moving projectile; its origin I conceive to be as follows:

Imagine any particle projected along a horizontal plane without friction; then we know, from what has been more fully explained in the preceding pages, that this particle will move along this same plane with a motion which is uniform and perpetual, provided the plane has no limits. But if the plane is limited and elevated, then the moving particle, which we imagine to be a heavy one, will on passing over the edge of the plane acquire, in addition to its previous uniform and perpetual motion, a downward propensity due to its own weight; so that the resulting motion which I call projection [*projectio*], is compounded of one which is uniform and horizontal and of another which is vertical and naturally accelerated. We now proceed to demonstrate

demonstrate some of its properties, the first of which is as follows:

[269]
THEOREM I, PROPOSITION I

A projectile which is carried by a uniform horizontal motion compounded with a naturally accelerated vertical motion describes a path which is a semi-parabola.

SAGR. Here, Salviati, it will be necessary to stop a little while for my sake and, I believe, also for the benefit of Simplicio; for it so happens that I have not gone very far in my study of Apollonius and am merely aware of the fact that he treats of the parabola and other conic sections, without an understanding of which I hardly think one will be able to follow the proof of other propositions depending upon them. Since even in this first beautiful theorem the author finds it necessary to prove that the path of a projectile is a parabola, and since, as I imagine, we shall have to deal with only this kind of curves, it will be absolutely necessary to have a thorough acquaintance, if not with all the properties which Apollonius has demonstrated for these figures, at least with those which are needed for the present treatment.

SALV. You are quite too modest, pretending ignorance of facts which not long ago you acknowledged as well known—I mean at the time when we were discussing the strength of materials and needed to use a certain theorem of Apollonius which gave you no trouble.

SAGR. I may have chanced to know it or may possibly have assumed it, so long as needed, for that discussion; but now when we have to follow all these demonstrations about such curves we ought not, as they say, to swallow it whole, and thus waste time and energy.

SIMP. Now even though Sagredo is, as I believe, well equipped for all his needs, I do not understand even the elementary terms; for although our philosophers have treated the motion of projectiles, I do not recall their having described the path of a projectile except to state in a general way that it is always a
curved

curved line, unless the projection be vertically upwards. But
[270]
if the little Euclid which I have learned since our previous discussion does not enable me to understand the demonstrations
which are to follow, then I shall be obliged to accept the theorems on faith without fully comprehending them.

SALV. On the contrary, I desire that you should understand
them from the Author himself, who, when he allowed me to see
this work of his, was good enough to prove for me two of the
principal properties of the parabola because I did not happen to
have at hand the books of Apollonius. These properties, which
are the only ones we shall need in the present discussion, he
proved in such a way that no prerequisite knowledge was required. These theorems are, indeed, given by Apollonius, but
after many preceding ones, to follow which would take a long
while. I wish to shorten our task by deriving the first property

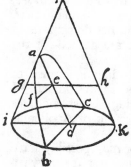

Fig. 106

purely and simply from the mode of gen-
eration of the parabola and proving the
second immediately from the first.

Beginning now with the first, imagine
a right cone, erected upon the circular
base *ibkc* with apex at *l*. The section of
this cone made by a plane drawn parallel
to the side *lk* is the curve which is called
a *parabola*. The base of this parabola *bc*
cuts at right angles the diameter *ik* of the
circle *ibkc*, and the axis *ad* is parallel to
the side *lk;* now having taken any point *f*
in the curve *bfa* draw the straight line *fe*
parallel to *bd;* then, I say, the square
of *bd* is to the square of *fe* in the same ratio as the axis *ad*
is to the portion *ae*. Through the point *e* pass a plane parallel
to the circle *ibkc*, producing in the cone a circular section whose
diameter is the line *geh*. Since *bd* is at right angles to *ik* in the
circle *ibk*, the square of *bd* is equal to the rectangle formed by *id*
and *dk;* so also in the upper circle which passes through the
points *gfh* the square of *fe* is equal to the rectangle formed by
ge

ge and *eh;* hence the square of *bd* is to the square of *fe* as the rectangle *id.dk* is to the rectangle *ge.eh.* And since the line *ed* is parallel to *hk,* the line *eh,* being parallel to *dk,* is equal to it; therefore the rectangle *id.dk* is to the rectangle *ge.eh* as *id* is to

[271]

ge, that is, as *da* is to *ae;* whence also the rectangle *id.dk* is to the rectangle *ge.eh,* that is, the square of *bd* is to the square of *fe,* as the axis *da* is to the portion *ae.*   Q. E. D.

The other proposition necessary for this discussion we demonstrate as follows. Let us draw a parabola whose axis *ca* is prolonged upwards to a point *d;* from any point *b* draw the line *bc* parallel to the base of the parabola; if now the point *d* is chosen

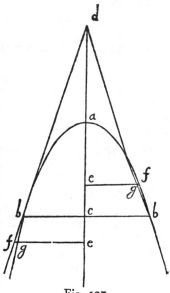

so that *da = ca,* then, I say, the straight line drawn through the points *b* and *d* will be tangent to the parabola at *b.* For imagine, if possible, that this line cuts the parabola above or that its prolongation cuts it below, and through any point *g* in it draw the straight line *fge.* And since the square of *fe* is greater than the square of *ge,* the square of *fe* will bear a greater ratio to the square of *bc* than the square of *ge* to that of *bc;* and since, by the preceding proposition, the square of *fe* is to that of *bc* as the line *ea* is to *ca,* it follows that the line *ea* will bear to the line *ca* a greater ratio than the square of *ge* to that of *bc,* or, than the square of *ed* to that of *cd* (the sides of the triangles *deg* and *dcb* being proportional).

Fig. 107

But the line *ea* is to *ca,* or *da,* in the same ratio as four times the rectangle *ea.ad* is to four times the square of *ad,* or, what is the same, the square of *cd,* since this is four times the square of *ad;* hence four times the rectangle *ea.ad* bears to the square of *cd*

a

a greater ratio than the square of *ed* to the square of *cd;* but that would make four times the rectangle *ea.ad* greater than the square of *ed;* which is false, the fact being just the opposite, because the two portions *ea* and *ad* of the line *ed* are not equal. Therefore the line *db* touches the parabola without cutting it.   Q. E. D.

SIMP. Your demonstration proceeds too rapidly and, it seems to me, you keep on assuming that all of Euclid's theorems are

[272]

as familiar and available to me as his first axioms, which is far from true. And now this fact which you spring upon us, that four times the rectangle *ea.ad* is less than the square of *de* because the two portions *ea* and *ad* of the line *de* are not equal brings me little composure of mind, but rather leaves me in suspense.

SALV. Indeed, all real mathematicians assume on the part of the reader perfect familiarity with at least the elements of Euclid; and here it is necessary in your case only to recall a proposition of the Second Book in which he proves that when a line is cut into equal and also into two unequal parts, the rectangle formed on the unequal parts is less than that formed on the equal (i. e., less than the square on half the line), by an amount which is the square of the difference between the equal and unequal segments. From this it is clear that the square of the whole line which is equal to four times the square of the half is greater than four times the rectangle of the unequal parts. In order to understand the following portions of this treatise it will be necessary to keep in mind the two elemental theorems from conic sections which we have just demonstrated; and these two theorems are indeed the only ones which the Author uses. We can now resume the text and see how he demonstrates his first proposition in which he shows that a body falling with a motion compounded of a uniform horizontal and a naturally accelerated [*naturale descendente*] one describes a semi-parabola.

Let us imagine an elevated horizontal line or plane *ab* along which a body moves with uniform speed from *a* to *b*. Suppose this

this plane to end abruptly at b; then at this point the body will, on account of its weight, acquire also a natural motion downwards along the perpendicular bn. Draw the line be along the plane ba to represent the flow, or measure, of time; divide this line into a number of segments, bc, cd, de, representing equal intervals of time; from the points b, c, d, e, let fall lines which are

parallel to the perpendicular bn. On the first of these lay off any distance ci, on the second a distance four times as long, df; on

[273]

the third, one nine times as long, eh; and so on, in proportion to the squares of cb, db, eb, or, we may say, in

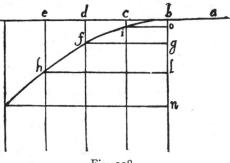

Fig. 108

the squared ratio of these same lines. Accordingly we see that while the body moves from b to c with uniform speed, it also falls perpendicularly through the distance ci, and at the end of the time-interval bc finds itself at the point i. In like manner at the end of the time-interval bd, which is the double of bc, the vertical fall will be four times the first distance ci; for it has been shown in a previous discussion that the distance traversed by a freely falling body varies as the square of the time; in like manner the space eh traversed during the time be will be nine times ci; thus it is evident that the distances eh, df, ci will be to one another as the squares of the lines be, bd, bc. Now from the points i, f, h draw the straight lines io, fg, hl parallel to be; these lines hl, fg, io are equal to eb, db and cb, respectively; so also are the lines bo, bg, bl respectively equal to ci, df, and eh. The square of hl is to that of fg as the line lb is to bg; and the square of fg is to that of io as gb is to bo; therefore the points i, f, h, lie on one and the same parabola. In like manner it may be shown that, if we take equal time-intervals of any size whatever, and if we imagine the particle to be carried by a similar compound motion, the

the positions of this particle, at the ends of these time-intervals, will lie on one and the same parabola.   Q. E. D.

Salv. This conclusion follows from the converse of the first of the two propositions given above. For, having drawn a parabola through the points *b* and *h*, any other two points, *f* and *i*, not falling on the parabola must lie either within or without; consequently the line *fg* is either longer or shorter than the line which terminates on the parabola. Therefore the square of *hl* will not bear to the square of *fg* the same ratio as the line *lb* to *bg*, but a greater or smaller; the fact is, however, that the square of *hl* *does* bear this same ratio to the square of *fg*. Hence the point *f* does lie on the parabola, and so do all the others.

Sagr. One cannot deny that the argument is new, subtle and conclusive, resting as it does upon this hypothesis, namely, that the horizontal motion remains uniform, that the vertical motion continues to be accelerated downwards in proportion to the square of the time, and that such motions and velocities as these combine without altering, disturbing, or hindering each other,* so that as the motion proceeds the path of the projectile does not change into a different curve: but this, in my opinion,

[274]

is impossible. For the axis of the parabola along which we imagine the natural motion of a falling body to take place stands perpendicular to a horizontal surface and ends at the center of the earth; and since the parabola deviates more and more from its axis no projectile can ever reach the center of the earth or, if it does, as seems necessary, then the path of the projectile must transform itself into some other curve very different from the parabola.

Simp. To these difficulties, I may add others. One of these is that we suppose the horizontal plane, which slopes neither up nor down, to be represented by a straight line as if each point on this line were equally distant from the center, which is not the case; for as one starts from the middle [of the line] and goes toward either end, he departs farther and farther from the center [of the earth] and is therefore constantly going uphill. Whence it follows that the motion cannot remain uniform

---

* A very near approach to Newton's Second Law of Motion. [*Trans.*]

through any distance whatever, but must continually diminish. Besides, I do not see how it is possible to avoid the resistance of the medium which must destroy the uniformity of the horizontal motion and change the law of acceleration of falling bodies. These various difficulties render it highly improbable that a result derived from such unreliable hypotheses should hold true in practice.

Salv. All these difficulties and objections which you urge are so well founded that it is impossible to remove them; and, as for me, I am ready to admit them all, which indeed I think our Author would also do. I grant that these conclusions proved in the abstract will be different when applied in the concrete and will be fallacious to this extent, that neither will the horizontal motion be uniform nor the natural acceleration be in the ratio assumed, nor the path of the projectile a parabola, etc. But, on the other hand, I ask you not to begrudge our Author that which other eminent men have assumed even if not strictly true. The authority of Archimedes alone will satisfy everybody. In his Mechanics and in his first quadrature of the parabola he takes for granted that the beam of a balance or steelyard is a straight line, every point of which is equidistant from the common center of all heavy bodies, and that the cords by which heavy bodies are suspended are parallel to each other.

Some consider this assumption permissible because, in practice, our instruments and the distances involved are so small in comparison with the enormous distance from the center of the earth that we may consider a minute of arc on a great circle as a straight line, and may regard the perpendiculars let fall from its two extremities as parallel. For if in actual practice one had to

[275]

consider such small quantities, it would be necessary first of all to criticise the architects who presume, by use of a plumbline, to erect high towers with parallel sides. I may add that, in all their discussions, Archimedes and the others considered themselves as located at an infinite distance from the center of the earth, in which case their assumptions were not false, and therefore their conclusions were absolutely correct. When we wish

wish to apply our proven conclusions to distances which, though finite, are very large, it is necessary for us to infer, on the basis of demonstrated truth, what correction is to be made for the fact that our distance from the center of the earth is not really infinite, but merely very great in comparison with the small dimensions of our apparatus. The largest of these will be the range of our projectiles—and even here we need consider only the artillery—which, however great, will never exceed four of those miles of which as many thousand separate us from the center of the earth; and since these paths terminate upon the surface of the earth only very slight changes can take place in their parabolic figure which, it is conceded, would be greatly altered if they terminated at the center of the earth.

As to the perturbation arising from the resistance of the medium this is more considerable and does not, on account of its manifold forms, submit to fixed laws and exact description. Thus if we consider only the resistance which the air offers to the motions studied by us, we shall see that it disturbs them all and disturbs them in an infinite variety of ways corresponding to the infinite variety in the form, weight, and velocity of the projectiles. For as to velocity, the greater this is, the greater will be the resistance offered by the air; a resistance which will be greater as the moving bodies become less dense [*men gravi*]. So that although the falling body ought to be displaced [*andare accelerandosi*] in proportion to the square of the duration of its motion, yet no matter how heavy the body, if it falls from a very considerable height, the resistance of the air will be such as to prevent any increase in speed and will render the motion

[276]

uniform; and in proportion as the moving body is less dense [*men grave*] this uniformity will be so much the more quickly attained and after a shorter fall. Even horizontal motion which, if no impediment were offered, would be uniform and constant is altered by the resistance of the air and finally ceases; and here again the less dense [*piu leggiero*] the body the quicker the process. Of these properties [*accidenti*] of weight, of velocity, and also of form [*figura*], infinite in number, it is not possible to give

give any exact description; hence, in order to handle this matter
in a scientific way, it is necessary to cut loose from these difficulties; and having discovered and demonstrated the theorems, in
the case of no resistance, to use them and apply them with such
limitations as experience will teach. And the advantage of this
method will not be small; for the material and shape of the
projectile may be chosen, as dense and round as possible, so
that it will encounter the least resistance in the medium. Nor
will the spaces and velocities in general be so great but that we
shall be easily able to correct them with precision.

In the case of those projectiles which we use, made of dense
[*grave*] material and round in shape, or of lighter material and
cylindrical in shape, such as arrows, thrown from a sling or
crossbow, the deviation from an exact parabolic path is quite
insensible. Indeed, if you will allow me a little greater liberty,
I can show you, by two experiments, that the dimensions of our
apparatus are so small that these external and incidental resistances, among which that of the medium is the most considerable, are scarcely observable.

I now proceed to the consideration of motions through the
air, since it is with these that we are now especially concerned;
the resistance of the air exhibits itself in two ways: first by
offering greater impedance to less dense than to very dense
bodies, and secondly by offering greater resistance to a body in
rapid motion than to the same body in slow motion.

Regarding the first of these, consider the case of two balls
having the same dimensions, but one weighing ten or twelve
times as much as the other; one, say, of lead, the other of oak,
both allowed to fall from an elevation of 150 or 200 cubits.

Experiment shows that they will reach the earth with slight
difference in speed, showing us that in both cases the retardation
caused by the air is small; for if both balls start at the same
moment and at the same elevation, and if the leaden one be
slightly retarded and the wooden one greatly retarded, then the
former ought to reach the earth a considerable distance in
advance of the latter, since it is ten times as heavy. But this

[277]

does

does not happen; indeed, the gain in distance of one over the other does not amount to the hundredth part of the entire fall. And in the case of a ball of stone weighing only a third or half as much as one of lead, the difference in their times of reaching the earth will be scarcely noticeable. Now since the speed [*impeto*] acquired by a leaden ball in falling from a height of 200 cubits is so great that if the motion remained uniform the ball would, in an interval of time equal to that of the fall, traverse 400 cubits, and since this speed is so considerable in comparison with those which, by use of bows or other machines except fire arms, we are able to give to our projectiles, it follows that we may, without sensible error, regard as absolutely true those propositions which we are about to prove without considering the resistance of the medium.

Passing now to the second case, where we have to show that the resistance of the air for a rapidly moving body is not very much greater than for one moving slowly, ample proof is given by the following experiment. Attach to two threads of equal length—say four or five yards—two equal leaden balls and suspend them from the ceiling; now pull them aside from the perpendicular, the one through 80 or more degrees, the other through not more than four or five degrees; so that, when set free, the one falls, passes through the perpendicular, and describes large but slowly decreasing arcs of 160, 150, 140 degrees, etc.; the other swinging through small and also slowly diminishing arcs of 10, 8, 6, degrees, etc.

In the first place it must be remarked that one pendulum passes through its arcs of 180°, 160°, etc., in the same time that the other swings through its 10°, 8°, etc., from which it follows that the speed of the first ball is 16 and 18 times greater than that of the second. Accordingly, if the air offers more resistance to the high speed than to the low, the frequency of vibration in the large arcs of 180° or 160°, etc., ought to be less than in the small arcs of 10°, 8°, 4°, etc., and even less than in arcs of 2°, or 1°; but this prediction is not verified by experiment; because if two persons start to count the vibrations, the one the large, the other the small, they will discover that after counting tens and

and even hundreds they will not differ by a single vibration, not even by a fraction of one.

[278]

This observation justifies the two following propositions, namely, that vibrations of very large and very small amplitude all occupy the same time and that the resistance of the air does not affect motions of high speed more than those of low speed, contrary to the opinion hitherto generally entertained.

SAGR. On the contrary, since we cannot deny that the air hinders both of these motions, both becoming slower and finally vanishing, we have to admit that the retardation occurs in the same proportion in each case. But how? How, indeed, could the resistance offered to the one body be greater than that offered to the other except by the impartation of more momentum and speed [*impeto e velocità*] to the fast body than to the slow? And if this is so the speed with which a body moves is at once the cause and measure [*cagione e misura*] of the resistance which it meets. Therefore, all motions, fast or slow, are hindered and diminished in the same proportion; a result, it seems to me, of no small importance.

SALV. We are able, therefore, in this second case to say that the errors, neglecting those which are accidental, in the results which we are about to demonstrate are small in the case of our machines where the velocities employed are mostly very great and the distances negligible in comparison with the semi-diameter of the earth or one of its great circles.

SIMP. I would like to hear your reason for putting the projectiles of fire arms, i. e., those using powder, in a different class from the projectiles employed in bows, slings, and crossbows, on the ground of their not being equally subject to change and resistance from the air.

SALV. I am led to this view by the excessive and, so to speak, supernatural violence with which such projectiles are launched; for, indeed, it appears to me that without exaggeration one might say that the speed of a ball fired either from a musket or from a piece of ordnance is supernatural. For if such a ball be allowed to fall from some great elevation its speed will, owing to the
resistance

resistance of the air, not go on increasing indefinitely; that which happens to bodies of small density in falling through short distances—I mean the reduction of their motion to uniformity—will also happen to a ball of iron or lead after it has fallen a few thousand cubits; this terminal or final speed [*terminata velocità*] is the maximum which such a heavy body can naturally acquire

[279]

in falling through the air. This speed I estimate to be much smaller than that impressed upon the ball by the burning powder.

An appropriate experiment will serve to demonstrate this fact. From a height of one hundred or more cubits fire a gun [*archibuso*] loaded with a lead bullet, vertically downwards upon a stone pavement; with the same gun shoot against a similar stone from a distance of one or two cubits, and observe which of the two balls is the more flattened. Now if the ball which has come from the greater elevation is found to be the less flattened of the two, this will show that the air has hindered and diminished the speed initially imparted to the bullet by the powder, and that the air will not permit a bullet to acquire so great a speed, no matter from what height it falls; for if the speed impressed upon the ball by the fire does not exceed that acquired by it in falling freely [*naturalmente*] then its downward blow ought to be greater rather than less.

This experiment I have not performed, but I am of the opinion that a musket-ball or cannon-shot, falling from a height as great as you please, will not deliver so strong a blow as it would if fired into a wall only a few cubits distant, i. e., at such a short range that the splitting or rending of the air will not be sufficient to rob the shot of that excess of supernatural violence given it by the powder.

The enormous momentum [*impeto*] of these violent shots may cause some deformation of the trajectory, making the beginning of the parabola flatter and less curved than the end; but, so far as our Author is concerned, this is a matter of small consequence in practical operations, the main one of which is the preparation of a table of ranges for shots of high elevation, giving the distance

tance attained by the ball as a function of the angle of elevation; and since shots of this kind are fired from mortars [*mortari*]
using small charges and imparting no supernatural momentum
[*impeto sopranaturale*] they follow their prescribed paths very
exactly.

But now let us proceed with the discussion in which the
Author invites us to the study and investigation of the motion
of a body [*impeto del mobile*] when that motion is compounded of
two others; and first the case in which the two are uniform, the
one horizontal, the other vertical.

[280]
THEOREM II, PROPOSITION II

When the motion of a body is the resultant of two uniform
motions, one horizontal, the other perpendicular, the square
of the resultant momentum is equal to the sum of the
squares of the two component momenta.*

Let us imagine any body urged by two uniform motions and
let *ab* represent the vertical displacement, while *bc* represents
the displacement which, in the same interval
of time, takes place in a horizontal direction. If then the distances *ab* and *bc* are
traversed, during the same time-interval,
with uniform motions the corresponding

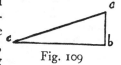
Fig. 109

momenta will be to each other as the distances *ab* and *bc* are to
each other; but the body which is urged by these two motions
describes the diagonal *ac*; its momentum is proportional to *ac*.
Also the square of *ac* is equal to the sum of the squares of *ab*
and *bc*. Hence the square of the resultant momentum is equal
to the sum of the squares of the two momenta *ab* and *bc*. Q. E. D.

SIMP. At this point there is just one slight difficulty which
needs to be cleared up; for it seems to me that the conclusion

* In the original this theorem reads as follows:
"*Si aliquod mobile duplici motu æquabili moveatur, nempe orizontali et
perpendiculari, impetus seu momentum lationis ex utroque motu com-
positæ erit potentia æqualis ambobus momentis priorum motuum.*"
For the justification of this translation of the word "*potentia*" and
of the use of the adjective "resultant" see p. 266 below. [*Trans.*]

just reached contradicts a previous proposition * in which it is claimed that the speed [*impeto*] of a body coming from *a* to *b* is equal to that in coming from *a* to *c;* while now you conclude that the speed [*impeto*] at *c* is greater than that at *b*.

SALV. Both propositions, Simplicio, are true, yet there is a great difference between them. Here we are speaking of a body urged by a single motion which is the resultant of two uniform motions, while there we were speaking of two bodies each urged with naturally accelerated motions, one along the vertical *ab* the other along the inclined plane *ac*. Besides the time-intervals were there not supposed to be equal, that along the incline *ac* being greater than that along the vertical *ab;* but the motions of which we now speak, those along *ab, bc, ac*, are uniform and simultaneous.

SIMP. Pardon me; I am satisfied; pray go on.

[281]

SALV. Our Author next undertakes to explain what happens when a body is urged by a motion compounded of one which is horizontal and uniform and of another which is vertical but naturally accelerated; from these two components results the path of a projectile, which is a parabola. The problem is to determine the speed [*impeto*] of the projectile at each point. With this purpose in view our Author sets forth as follows the manner, or rather the method, of measuring such speed [*impeto*] along the path which is taken by a heavy body starting from rest and falling with a naturally accelerated motion.

THEOREM III, PROPOSITION III

Let the motion take place along the line *ab*, starting from rest at *a*, and in this line choose any point *c*. Let *ac* represent the time, or the measure of the time, required for the body to fall through the space *ac;* let *ac* also represent the velocity [*impetus seu momentum*] at *c* acquired by a fall through the distance *ac*. In the line *ab* select any other point *b*. The problem now is to determine the velocity at *b* acquired by a body in falling through the distance *ab* and to express this in terms of the velocity at *c*, the measure of which is the length *ac*. Take

* See p. 169 above. [*Trans.*]

as a mean proportional between *ac* and *ab*. We shall prove
that the velocity at *b* is to that at *c* as the length *as* is to the
length *ac*. Draw the horizontal
line *cd*, having twice the length
of *ac*, and *be*, having twice the
length of *ba*. It then follows,
from the preceding theorems,
that a body falling through the
distance *ac*, and turned so as
to move along the horizontal *cd*

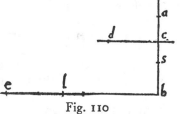

Fig. 110

with a uniform speed equal to that acquired on reaching *c*
[282]
will traverse the distance *cd* in the same interval of time as
that required to fall with accelerated motion from *a* to *c*. Likewise *be* will be traversed in the same time as *ba*. But the time
of descent through *ab* is *as;* hence the horizontal distance *be*
is also traversed in the time *as*. Take a point *l* such that the
time *as* is to the time *ac* as *be* is to *bl;* since the motion along
be is uniform, the distance *bl*, if traversed with the speed [*momentum celeritatis*] acquired at *b*, will occupy the time *ac;* but
in this same time-interval, *ac*, the distance *cd* is traversed with
the speed acquired in *c*. Now two speeds are to each other as
the distances traversed in equal intervals of time. Hence the
speed at *c* is to the speed at *b* as *cd* is to *bl*. But since *dc* is to
be as their halves, namely, as *ca* is to *ba*, and since *be* is to *bl*
as *ba* is to *sa;* it follows that *dc* is to *bl* as *ca* is to *sa*. In other
words, the speed at *c* is to that at *b* as *ca* is to *sa*, that is, as the
time of fall through *ab*.

The method of measuring the speed of a body along the direction of its fall is thus clear; the speed is assumed to increase
directly as the time.

But before we proceed further, since this discussion is to
deal with the motion compounded of a uniform horizontal one
and one accelerated vertically downwards—the path of a projectile, namely, a parabola—it is necessary that we define some
common standard by which we may estimate the velocity, or
momentum [*velocitatem, impetum seu momentum*] of both motions

tions; and since from the innumerable uniform velocities one only, and that not selected at random, is to be compounded with a velocity acquired by naturally accelerated motion, I can think of no simpler way of selecting and measuring this than to assume another of the same kind.* For the sake of clearness, draw the vertical line *ac* to meet the horizontal line *bc*. *Ac* is the height and *bc* the amplitude of the semi-parabola *ab*, which is the resultant of the two motions, one that of a body falling

[283]

from rest at *a*, through the distance *ac*, with naturally accelerated motion, the other a uniform motion along the horizontal *ad*. The speed acquired at *c* by a fall through the distance *ac* is determined by the height *ac*; for the speed of a body falling from the same elevation is always one and the same; but along the horizontal one may give a body an infinite number of uniform speeds. However, in order that I may select one out of this multitude and separate it from the rest in a perfectly definite manner, I will extend the height *ca* upwards to *e* just as far as is necessary and will call this distance *ae* the "sublimity." Imagine a body to fall from rest at *e*; it is clear that we may make its terminal speed at *a* the same as that with which the same body travels along the horizontal line *ad*; this speed will be such that, in the time of descent along *ea*, it will describe a horizontal distance twice the length of *ea*. This preliminary remark seems necessary.

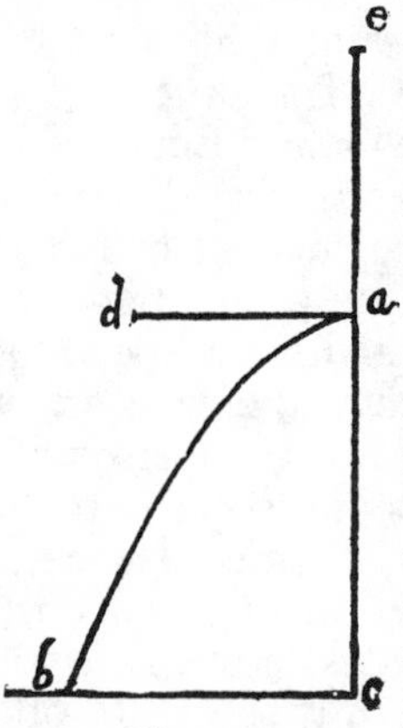

Fig. 111

The reader is reminded that above I have called the horizontal line *cb* the "amplitude" of the semi-parabola *ab*; the axis *ac* of this parabola, I have called its "altitude"; but the line *ea* the fall along which determines the horizontal speed I have called the "sublimity." These matters having been explained, I proceed with the demonstration.

* Galileo here proposes to employ as a standard of velocity the terminal speed of a body falling freely from a given height. [*Trans.*]

SAGR. Allow me, please, to interrupt in order that I may point out the beautiful agreement between this thought of the Author and the views of Plato concerning the origin of the various uniform speeds with which the heavenly bodies revolve. The latter chanced upon the idea that a body could not pass from rest to any given speed and maintain it uniformly except by passing through all the degrees of speed intermediate between the given speed and rest. Plato thought that God, after having created the heavenly bodies, assigned them the proper and uniform speeds with which they were forever to revolve; and that He made them start from rest and move over definite distances under a natural and rectilinear acceleration such as governs the motion of terrestrial bodies. He added that once these bodies had gained their proper and permanent speed, their rectilinear motion was converted into a circular one, the only

[284]

motion capable of maintaining uniformity, a motion in which the body revolves without either receding from or approaching its desired goal. This conception is truly worthy of Plato; and it is to be all the more highly prized since its underlying principles remained hidden until discovered by our Author who removed from them the mask and poetical dress and set forth the idea in correct historical perspective. In view of the fact that astronomical science furnishes us such complete information concerning the size of the planetary orbits, the distances of these bodies from their centers of revolution, and their velocities, I cannot help thinking that our Author (to whom this idea of Plato was not unknown) had some curiosity to discover whether or not a definite "sublimity" might be assigned to each planet, such that, if it were to start from rest at this particular height and to fall with naturally accelerated motion along a straight line, and were later to change the speed thus acquired into uniform motion, the size of its orbit and its period of revolution would be those actually observed.

SALV. I think I remember his having told me that he once made the computation and found a satisfactory correspondence with observation. But he did not wish to speak of it, lest in
view

view of the odium which his many new discoveries had already brought upon him, this might be adding fuel to the fire. But if any one desires such information he can obtain it for himself from the theory set forth in the present treatment.

We now proceed with the matter in hand, which is to prove:

PROBLEM I, PROPOSITION IV

To determine the momentum of a projectile at each particular point in its given parabolic path.

Let *bec* be the semi-parabola whose amplitude is *cd* and whose height is *db*, which latter extended upwards cuts the tangent of the parabola *ca* in *a*. Through the vertex draw the horizontal line *bi* parallel to *cd*. Now if the amplitude *cd* is equal to the entire height *da*, then *bi* will be equal to *ba* and also to *bd;* and if we take *ab* as the measure of the time required for fall through the distance *ab* and also of the momentum acquired at *b* in consequence of its fall from rest at *a*, then if we turn into a horizontal direction the momentum acquired by fall through *ab* [*impetum ab*] the space traversed in the same interval of time will be represented by *dc* which is twice *bi*. But a body which falls from rest at *b* along the line *bd* will during the same time-interval fall through the height of the parabola

[285]

bd. Hence a body falling from rest at *a*, turned into a horizontal direction with the speed *ab* will traverse a space equal to *dc*. Now if one superposes upon this motion a fall along *bd*, traversing the height *bd* while the parabola *bc* is described, then the momentum of the body at the terminal point *c* is the resultant of a uniform horizontal momentum, whose value is represented by *ab*, and of another momentum acquired by fall from *b* to the terminal point *d* or *c;* these two momenta are equal. If, therefore, we take *ab* to be the measure of one of these momenta, say, the uniform horizontal one, then *bi*, which is equal to *bd*, will represent the momentum acquired at *d* or *c;* and *ia* will represent the resultant of these two momenta, that is, the total momentum with which the projectile, travelling along the parabola, strikes at *c*.

With

With this in mind let us take any point on the parabola, say *e*, and determine the momentum with which the projectile passes that point. Draw the horizontal *ef* and take *bg* a mean proportional between *bd* and *bf*. Now since *ab*, or *bd*, is as-sumed to be the measure of the time and of the momentum [*momentum velocitatis*] acquired by fall-ing from rest at *b* through the dis-tance *bd*, it follows that *bg* will measure the time and also the momentum [*impetus*] acquired at *f* by fall from *b*. If therefore we lay off *bo*, equal to *bg*, the diagonal line joining *a* and *o* will represent the momentum at the point *e;* because the length *ab* has been assumed to represent the momentum at *b*

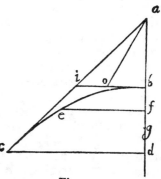

Fig. 112

which, after diversion into a horizontal direction, remains con-stant; and because *bo* measures the momentum at *f* or *e*, ac-quired by fall, from rest at *b*, through the height *bf*. But the square of *ao* equals the sum of the squares of *ab* and *bo*. Hence the theorem sought.

SAGR. The manner in which you compound these different momenta to obtain their resultant strikes me as so novel that my mind is left in no small confusion. I do not refer to the composition of two uniform motions, even when unequal, and when one takes place along a horizontal, the other along a vertical direction; because in this case I am thoroughly con-vinced that the resultant is a motion whose square is equal to the sum of the squares of the two components. The confusion arises when one undertakes to compound a uniform horizontal motion with a vertical one which is naturally accelerated. I trust, therefore, we may pursue this discussion more at length.

[286]

SIMP. And I need this even more than you since I am not yet as clear in my mind as I ought to be concerning those funda-mental propositions upon which the others rest. Even in the case

case of the two uniform motions, one horizontal, the other perpendicular, I wish to understand better the manner in which you obtain the resultant from the components. Now, Salviati, you understand what we need and what we desire.

SALV. Your request is altogether reasonable and I will see whether my long consideration of these matters will enable me to make them clear to you. But you must excuse me if in the explanation I repeat many things already said by the Author.

Concerning motions and their velocities or momenta [*movimenti e lor velocità o impeti*] whether uniform or naturally accelerated, one cannot speak definitely until he has established a measure for such velocities and also for time. As for time we have the already widely adopted hours, first minutes and second minutes. So for velocities, just as for intervals of time, there is need of a common standard which shall be understood and accepted by everyone, and which shall be the same for all. As has already been stated, the Author considers the velocity of a freely falling body adapted to this purpose, since this velocity increases according to the same law in all parts of the world; thus for instance the speed acquired by a leaden ball of a pound weight starting from rest and falling vertically through the height of, say, a spear's length is the same in all places; it is therefore excellently adapted for representing the momentum [*impeto*] acquired in the case of natural fall.

It still remains for us to discover a method of measuring momentum in the case of uniform motion in such a way that all who discuss the subject will form the same conception of its size and velocity [*grandezza e velocità*]. This will prevent one person from imagining it larger, another smaller, than it really is; so that in the composition of a given uniform motion with one which is accelerated different men may not obtain different values for the resultant. In order to determine and represent such a momentum and particular speed [*impeto e velocità particolare*] our Author has found no better method than to use the momentum acquired by a body in naturally accelerated motion.

[287]

The speed of a body which has in this manner acquired any
momentum

momentum whatever will, when converted into uniform motion, retain precisely such a speed as, during a time-interval equal to that of the fall, will carry the body through a distance equal to twice that of the fall. But since this matter is one which is fundamental in our discussion it is well that we make it perfectly clear by means of some particular example.

Let us consider the speed and momentum acquired by a body falling through the height, say, of a spear [*picca*] as a standard which we may use in the measurement of other speeds and momenta as occasion demands; assume for instance that the time of such a fall is four seconds [*minuti secondi d'ora*]; now in order to measure the speed acquired from a fall through any other height, whether greater or less, one must not conclude that these speeds bear to one another the same ratio as the heights of fall; for instance, it is not true that a fall through four times a given height confers a speed four times as great as that acquired by descent through the given height; because the speed of a naturally accelerated motion does not vary in proportion to the time. As has been shown above, the ratio of the spaces is equal to the square of the ratio of the times.

If, then, as is often done for the sake of brevity, we take the same limited straight line as the measure of the speed, and of the time, and also of the space traversed during that time, it follows that the duration of fall and the speed acquired by the same body in passing over any other distance, is not represented by this second distance, but by a mean proportional between the two distances. This I can better illustrate by an example. In the vertical line *ac*, lay off the portion *ab* to represent the distance traversed by a body falling freely with accelerated motion: the time of fall may be represented by any limited straight line, but for the sake of brevity, we shall represent it by the same length *ab;* this length may also be employed as a measure of the momentum and speed acquired during the motion; in short, let *ab* be a measure of the various physical quantities which enter this discussion.

Fig. 113

Having agreed arbitrarily upon *ab* as a measure of these three

three different quantities, namely, space, time, and momentum, our next task is to find the time required for fall through a
[288]
given vertical distance ac, also the momentum acquired at the terminal point c, both of which are to be expressed in terms of the time and momentum represented by ab. These two required quantities are obtained by laying off ad, a mean proportional between ab and ac; in other words, the time of fall from a to c is represented by ad on the same scale on which we agreed that the time of fall from a to b should be represented by ab. In like manner we may say that the momentum [*impeto o grado di velocità*] acquired at c is related to that acquired at b, in the same manner that the line ad is related to ab, since the velocity varies directly as the time, a conclusion, which although employed as a postulate in Proposition III, is here amplified by the Author.

This point being clear and well-established we pass to the consideration of the momentum [*impeto*] in the case of two compound motions, one of which is compounded of a uniform horizontal and a uniform vertical motion, while the other is compounded of a uniform horizontal and a naturally accelerated vertical motion. If both components are uniform, and one at right angles to the other, we have already seen that the square of the resultant is obtained by adding the squares of the components [p. 257] as will be clear from the following illustration.

Let us imagine a body to move along the vertical ab with a uniform momentum [*impeto*] of 3, and on reaching b to move

Fig. 114

toward c with a momentum [*velocità ed impeto*] of 4, so that during the same time-interval it will traverse 3 cubits along the vertical and 4 along the horizontal. But a particle which moves with the resultant velocity [*velocità*] will, in the same time, traverse the diagonal ac, whose length is not 7 cubits—the sum of ab (3) and bc (4)—but 5, which is *in potenza* equal to the sum of 3 and 4, that is, the squares of 3 and 4 when added make 25, which is the square of ac, and is equal to the sum of the squares
of

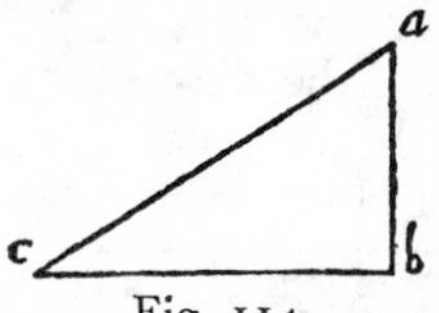

of *ab* and *bc*. Hence *ac* is represented by the side—or we may say the root—of a square whose area is 25, namely 5.

As a fixed and certain rule for obtaining the momentum which
[289]
results from two uniform momenta, one vertical, the other horizontal, we have therefore the following: take the square of each, add these together, and extract the square root of the sum, which will be the momentum resulting from the two. Thus, in the above example, the body which in virtue of its vertical motion would strike the horizontal plane with a momentum [*forza*] of 3, would owing to its horizontal motion alone strike at *c* with a momentum of 4; but if the body strikes with a momentum which is the resultant of these two, its blow will be that of a body moving with a momentum [*velocità e forza*] of 5; and such a blow will be the same at all points of the diagonal *ac*, since its components are always the same and never increase or diminish.

Let us now pass to the consideration of a uniform horizontal motion compounded with the vertical motion of a freely falling body starting from rest. It is at once clear that the diagonal which represents the motion compounded of these two is not a straight line, but, as has been demonstrated, a semi-parabola, in which the momentum [*impeto*] is always increasing because the speed [*velocità*] of the vertical component is always increasing. Wherefore, to determine the momentum [*impeto*] at any given point in the parabolic diagonal, it is necessary first to fix upon the uniform horizontal momentum [*impeto*] and then, treating the body as one falling freely, to find the vertical momentum at the given point; this latter can be determined only by taking into account the duration of fall, a consideration which does not enter into the composition of two uniform motions where the velocities and momenta are always the same; but here where one of the component motions has an initial value of zero and increases its speed [*velocità*] in direct proportion to the time, it follows that the time must determine the speed [*velocità*] at the assigned point. It only remains to obtain the momentum resulting from these two components (as in the case of uniform motions) by placing the square of the resultant equal
to

to the sum of the squares of the two components. But here
again it is better to illustrate by means of an example.

On the vertical *ac* lay off any portion *ab* which we shall employ as a measure of the space traversed by a body falling freely
along the perpendicular, likewise as a measure of the time and
also of the speed [*grado di velocità*] or, we may say, of the momenta [*impeti*].   It is at once clear that if the momentum of a
[290]
body at *b*, after having fallen from rest at *a*, be diverted along
the horizontal direction *bd*, with uniform motion, its speed will
be such that, during the time-interval *ab*, it will traverse a
distance which is represented by the line *bd* and which is twice as

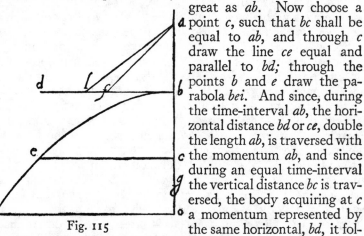

great as *ab*. Now choose a
point *c*, such that *bc* shall be
equal to *ab*, and through *c*
draw the line *ce* equal and
parallel to *bd;* through the
points *b* and *e* draw the parabola *bei*. And since, during
the time-interval *ab*, the horizontal distance *bd* or *ce*, double
the length *ab*, is traversed with
the momentum *ab*, and since
during an equal time-interval
the vertical distance *bc* is traversed, the body acquiring at *c*
a momentum represented by
the same horizontal, *bd*, it fol-

Fig. 115

lows that during the time *ab* the body will pass from *b* to *e* along
the parabola *be*, and will reach *e* with a momentum compounded
of two momenta each equal to *ab*. And since one of these is
horizontal and the other vertical, the square of the resultant momentum is equal to the sum of the squares of these two components, i. e., equal to twice either one of them.

Therefore, if we lay off the distance *bf*, equal to *ba*, and draw
the diagonal *af*, it follows that the momentum [*impeto e percossa*] at *e* will exceed that of a body at *b* after having fallen from

a,

a, or what is the same thing, will exceed the horizontal momentum [*percossa dell'impeto*] along *bd*, in the ratio of *af* to *ab*.

Suppose now we choose for the height of fall a distance *bo* which is not equal to but greater than *ab*, and suppose that *bg* represents a mean proportional between *ba* and *bo;* then, still retaining *ba* as a measure of the distance fallen through, from rest at *a*, to *b*, also as a measure of the time and of the momentum which the falling body acquires at *b*, it follows that *bg* will be the measure of the time and also of the momentum which the body acquires in falling from *b* to *o*. Likewise just as the momentum *ab* during the time *ab* carried the body a distance along the horizontal equal to twice *ab*, so now, during the time-interval *bg*, the body will be carried in a horizontal direction through a distance which is greater in the ratio of *bg* to *ba*. Lay off *lb* equal to *bg* and draw the diagonal *al*, from which we have a quantity compounded of two velocities [*impeti*] one horizontal, the other vertical; these determine the parabola. The horizontal and uniform velocity is that acquired at *b* in falling from *a;* the other is that acquired at *o*, or, we may say, at *i*, by a body falling through the distance *bo*, during a time measured by the line *bg*,

[291]

which line *bg* also represents the momentum of the body. And in like manner we may, by taking a mean proportional between the two heights, determine the momentum [*impeto*] at the extreme end of the parabola where the height is less than the sublimity *ab;* this mean proportional is to be drawn along the horizontal in place of *bf*, and also another diagonal in place of *af*, which diagonal will represent the momentum at the extreme end of the parabola.

To what has hitherto been said concerning the momenta, blows or shocks of projectiles, we must add another very important consideration; to determine the force and energy of the shock [*forza ed energia della percossa*] it is not sufficient to consider only the speed of the projectiles, but we must also take into account the nature and condition of the target which, in no small degree, determines the efficiency of the blow. First of all it is well known that the target suffers violence from the speed [*velocità*]

[*velocità*] of the projectile in proportion as it partly or entirely stops the motion; because if the blow falls upon an object which yields to the impulse [*velocità del percuziente*] without resistance such a blow will be of no effect; likewise when one attacks his enemy with a spear and overtakes him at an instant when he is fleeing with equal speed there will be no blow but merely a harmless touch. But if the shock falls upon an object which yields only in part then the blow will not have its full effect, but the damage will be in proportion to the excess of the speed of the projectile over that of the receding body; thus, for example, if the shot reaches the target with a speed of 10 while the latter recedes with a speed of 4, the momentum and shock [*impeto e percossa*] will be represented by 6. Finally the blow will be a maximum, in so far as the projectile is concerned, when the target does not recede at all but if possible completely resists and stops the motion of the projectile. I have said *in so far as the projectile is concerned* because if the target should approach the projectile the shock of collision [*colpo e l'incontro*] would be greater in proportion as the sum of the two speeds is greater than that of the projectile alone.

Moreover it is to be observed that the amount of yielding in the target depends not only upon the quality of the material, as regards hardness, whether it be of iron, lead, wool, etc., but

[292]

also upon its position. If the position is such that the shot strikes it at right angles, the momentum imparted by the blow [*impeto del colpo*] will be a maximum; but if the motion be oblique, that is to say slanting, the blow will be weaker; and more and more so in proportion to the obliquity; for, no matter how hard the material of the target thus situated, the entire momentum [*impeto e moto*] of the shot will not be spent and stopped; the projectile will slide by and will, to some extent, continue its motion along the surface of the opposing body.

All that has been said above concerning the amount of momentum in the projectile at the extremity of the parabola must be understood to refer to a blow received on a line at right angles to this parabola or along the tangent to the parabola at the given

point

point; for, even though the motion has two components, one horizontal, the other vertical, neither will the momentum along the horizontal nor that upon a plane perpendicular to the horizontal be a maximum, since each of these will be received obliquely.

SAGR. Your having mentioned these blows and shocks recalls to my mind a problem, or rather a question, in mechanics of which no author has given a solution or said anything which diminishes my astonishment or even partly relieves my mind.

My difficulty and surprise consist in not being able to see whence and upon what principle is derived the energy and immense force [*energia e forza immensa*] which makes its appearance in a blow; for instance we see the simple blow of a hammer, weighing not more than 8 or 10 lbs., overcoming resistances which, without a blow, would not yield to the weight of a body producing impetus by pressure alone, even though that body weighed many hundreds of pounds. I would like to discover a method of measuring the force [*forza*] of such a percussion. I can hardly think it infinite, but incline rather to the view that it has its limit and can be counterbalanced and measured by other forces, such as weights, or by levers or screws or other mechanical instruments which are used to multiply forces in a manner which I satisfactorily understand.

SALV. You are not alone in your surprise at this effect or in obscurity as to the cause of this remarkable property. I studied this matter myself for a while in vain; but my confusion merely increased until finally meeting our Academician I received from

[293]

him great consolation. First he told me that he also had for a long time been groping in the dark; but later he said that, after having spent some thousands of hours in speculating and contemplating thereon, he had arrived at some notions which are far removed from our earlier ideas and which are remarkable for their novelty. And since now I know that you would gladly hear what these novel ideas are I shall not wait for you to ask but promise that, as soon as our discussion of projectiles is completed, I will explain all these fantasies, or if you please,

vagaries

vagaries, as far as I can recall them from the words of our Academician. In the meantime we proceed with the propositions of the author.

PROPOSITION V, PROBLEM

Having given a parabola, find the point, in its axis extended upwards, from which a particle must fall in order to describe this same parabola.

Let *ab* be the given parabola, *hb* its amplitude, and *he* its axis extended. The problem is to find the point *e* from which a body must fall in order that, after the momentum which it acquires at *a* has been diverted into a horizontal direction, it will describe the parabola *ab*. Draw the horizontal *ag*, parallel to *bh*, and

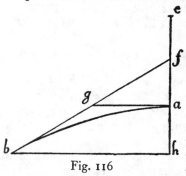

having laid off *af* equal to *ah*, draw the straight line *bf* which will be a tangent to the parabola at *b*, and will intersect the horizontal *ag* at *g*: choose *e* such that *ag* will be a mean proportional between *af* and *ae*. Now I say that *e* is the point above sought. That is, if a body falls from rest at this point *e*, and if the momentum acquired at the

Fig. 116

point *a* be diverted into a horizontal direction, and compounded with the momentum acquired at *h* in falling from rest at *a*, then the body will describe the parabola *ab*. For if we understand *ea* to be the measure of the time of fall from *e* to *a*, and also of the momentum acquired at *a*, then *ag* (which is a mean proportional between *ea* and *af*) will represent the time and momentum of fall from *f* to *a* or, what is the same thing, from *a* to *h*; and since a body falling from *e*, during the time *ea*, will, owing to the momentum acquired at *a*, traverse at uniform speed a horizontal distance which is twice *ea*, it follows that, the body will if impelled by the same momentum, during the time-interval *ag* traverse a distance equal to twice *ag* which is the half of *bh*. This is true because,

in

in the case of uniform motion, the spaces traversed vary directly as the times. And likewise if the motion be vertical and start from rest, the body will describe the distance *ah* in the

[294]

time *ag*. Hence the amplitude *bh* and the altitude *ah* are traversed by a body in the same time. Therefore the parabola *ab* will be described by a body falling from the sublimity of *e*.

Q. E. F.

COROLLARY

Hence it follows that half the base, or amplitude, of the semi-parabola (which is one-quarter of the entire amplitude) is a mean proportional between its altitude and the sublimity from which a falling body will describe this same parabola.

PROPOSITION VI, PROBLEM

Given the sublimity and the altitude of a parabola, to find its amplitude.

Let the line *ac*, in which lie the given altitude *cb* and sublimity *ab*, be perpendicular to the horizontal line *cd*. The problem is to find the amplitude, along the horizontal *cd*, of the semi-parabola which is described with the sublimity *ba* and altitude *bc*. Lay off *cd* equal to twice the mean proportional between *cb* and *ba*. Then *cd* will be the amplitude sought, as is evident from the preceding proposition.

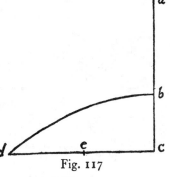

Fig. 117

THEOREM. PROPOSITION VII

If projectiles describe semi-parabolas of the same amplitude, the momentum required to describe that one whose amplitude is double its altitude is less than that required for any other.

Let

Let *bd* be a semi-parabola whose amplitude *cd* is double its altitude *cb;* on its axis extended upwards lay off *ba* equal to its altitude *bc*. Draw the line *ad* which will be a tangent to the parabola at *d* and will cut the horizontal line *be* at the point *e*, making *be* equal to *bc* and also to *ba*. It is evident that this parabola will be described by a projectile whose uniform horizontal momentum is that which it would acquire at *b* in falling from rest at *a* and whose naturally accelerated vertical momentum is that of the body falling to *c*, from rest at *b*. From this it follows

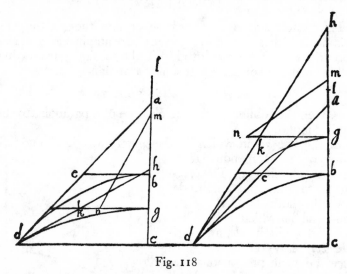

Fig. 118

that the momentum at the terminal point *d*, compounded of these two, is represented by the diagonal *ae*, whose square is equal to the sum of the squares of the two components. Now let *gd* be any other parabola whatever having the same amplitude *cd*, but whose altitude *cg* is either greater or less than the altitude *bc*. Let *hd* be the tangent cutting the horizontal through *g* at *k*. Select a point *l* such that *hg:gk = gk:gl*. Then from a preceding proposition [V], it follows that *gl* will be the height

[295]

height from which a body must fall in order to describe the parabola *gd*.

Let *gm* be a mean proportional between *ab* and *gl*; then *gm* will [Prop. IV] represent the time and momentum acquired at *g* by a fall from *l*; for *ab* has been assumed as a measure of both time and momentum. Again let *gn* be a mean proportional between *bc* and *cg*; it will then represent the time and momentum which the body acquires at *c* in falling from *g*. If now we join *m* and *n*, this line *mn* will represent the momentum at *d* of the projectile traversing the parabola *dg*; which momentum is, I say, greater than that of the projectile travelling along the parabola *bd* whose measure was given by *ae*. For since *gn* has been taken as a mean proportional between *bc* and *gc*; and since *bc* is equal to *be* and also to *kg* (each of them being the half of *dc*) it follows that $cg:gn = gn:gk$, and as *cg* or (*hg*) is to *gk* so is $\overline{ng}^2$ to $\overline{gk}^2$: but by construction $hg:gk = gk:gl$. Hence $\overline{ng}^2 : \overline{gk}^2 = gk:gl$. But $gk:gl = \overline{gk}^2 : \overline{gm}^2$, since *gm* is a mean proportional between *kg* and *gl*. Therefore the three squares *ng*, *kg*, *mg* form a continued proportion, $\overline{gn}^2 : \overline{gk}^2 = \overline{gk}^2 : \overline{gm}.^2$ And the sum of the two extremes which is equal to the square of *mn* is greater than twice the square of *gk*; but the square of *ae* is double the square of *gk*. Hence the square of *mn* is greater than the square of *ae* and the length *mn* is greater than the length *ae*.

Q. E. D.

[296]
COROLLARY

Conversely it is evident that less momentum will be required to send a projectile from the terminal point *d* along the parabola *bd* than along any other parabola having an elevation greater or less than that of the parabola *bd*, for which the tangent at *d* makes an angle of 45° with the horizontal. From which it follows that if projectiles are fired from the terminal point *d*, all having the same speed, but each having a different elevation, the maximum range, i. e., amplitude of the semi-parabola or of the entire parabola, will be obtained when the elevation is 45°: the other

other shots, fired at angles greater or less will have a shorter range.

SAGR. The force of rigid demonstrations such as occur only in mathematics fills me with wonder and delight. From accounts given by gunners, I was already aware of the fact that in the use of cannon and mortars, the maximum range, that is the one in which the shot goes farthest, is obtained when the elevation is 45° or, as they say, at the sixth point of the quadrant; but to understand why this happens far outweighs the mere information obtained by the testimony of others or even by repeated experiment.

SALV. What you say is very true. The knowledge of a single fact acquired through a discovery of its causes prepares the mind to understand and ascertain other facts without need of recourse to experiment, precisely as in the present case, where by argumentation alone the Author proves with certainty that the maximum range occurs when the elevation is 45°. He thus demonstrates what has perhaps never been observed in experience, namely, that of other shots those which exceed or fall short of 45° by equal amounts have equal ranges; so that if the balls have been fired one at an elevation of 7 points, the other at 5, they will strike the level at the same distance: the same is true if the shots are fired at 8 and at 4 points, at 9 and at 3, etc. Now let us hear the demonstration of this.

[297]
THEOREM. PROPOSITION VIII

The amplitudes of two parabolas described by projectiles fired with the same speed, but at angles of elevation which exceed and fall short of 45° by equal amounts, are equal to each other.

In the triangle *mcb* let the horizontal side *bc* and the vertical *cm*, which form a right angle at *c*, be equal to each other; then the angle *mbc* will be a semi-right angle; let the line *cm* be prolonged to *d*, such a point that the two angles at *b*, namely *mbe* and *mbd*, one above and the other below the diagonal *mb*, shall be equal. It is now to be proved that in the case of two parabolas described

described by two projectiles fired from b with the same speed, one at the angle of ebc, the other at the angle of dbc, their amplitudes will be equal. Now since the external angle bmc is equal to the sum of the internal angles mdb and dbm we may also equate to them the angle mbc; but if we replace the angle dbm by mbe, then this same angle mbc is equal to the two mbe and bdc: and if we subtract from each side of this equation the angle mbe, we have the remainder bdc equal to the remainder ebc. Hence the two triangles dcb and bce are similar. Bisect the straight lines dc and ec in the points h and f: and draw the lines hi and fg parallel to the horizontal cb, and choose l such that $dh:hi = ih:hl$. Then the triangle ihl will be similar to ihd, and also to the triangle egf; and since ih and gf are equal, each being half of bc, it follows that hl is equal to fe and also to fc; and if we add to each of these the common part fh, it will be seen that ch is equal to fl.

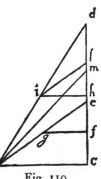

Fig. 119

Let us now imagine a parabola described through the points h and b whose altitude is hc and sublimity hl. Its amplitude will be cb which is double the length hi since hi is a mean proportional between dh (or ch) and hl. The line db is tangent to the parabola at b, since ch is equal to hd. If again we imagine a parabola described through the points f and b, with a sublimity fl and altitude fc, of which the mean proportional is fg, or one-half of cb, then, as before, will cb be the amplitude and the line eb a tangent at b; for ef and fc are equal.

[298]

But the two angles dbc and ebc, the angles of elevation, differ by equal amounts from a 45° angle. Hence follows the proposition.

THEOREM. PROPOSITION IX

The amplitudes of two parabolas are equal when their altitudes and sublimities are inversely proportional.

Let

Let the altitude *gf* of the parabola *fh* bear to the altitude *cb* of the parabola *bd* the same ratio which the sublimity *ba* bears to the sublimity *fe;* then I say the amplitude *hg* is equal to the amplitude *dc.* For since the first of these quantities, *gf*, bears to

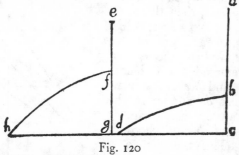

the second *cb* the same ratio which the third, *ba*, bears to the fourth *fe*, it follows that the area of the rectangle *gf.fe* is equal to that of the rectangle *cb.ba;* therefore squares which are equal to these rectangles are equal to each other.

Fig. 120

But [by Proposition VI] the square of half of *gh* is equal to the rectangle *gf.fe;* and the square of half of *cd* is equal to the rectangle *cb.ba.* Therefore these squares and their sides and the doubles of their sides are equal. But these last are the amplitudes *gh* and *cd.* Hence follows the proposition.

LEMMA FOR THE FOLLOWING PROPOSITION

If a straight line be cut at any point whatever and mean proportionals between this line and each of its parts be taken, the sum of the squares of these mean proportionals is equal to the square of the entire line.

Let the line *ab* be cut at *c.* Then I say that the square of the mean proportional between *ab* and *ac* plus the square of the mean proportional between *ab* and *cb* is equal to the square of the whole line *ab.* This is evident as soon as we describe a semicircle upon the entire line *ab*, erect a perpendicular *cd* at *c*, and draw *da* and *db.* For *da* is a mean proportional between *ab* and *ac* while

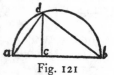

Fig. 121

[299]

db is a mean proportional between *ab* and *bc:* and since the angle *adb*, inscribed in a semicircle, is a right angle the sum of the

the squares of the lines *da* and *db* is equal to the square of the entire line *ab*. Hence follows the proposition.

THEOREM. PROPOSITION X

The momentum [*impetus seu momentum*] acquired by a particle at the terminal point of any semi-parabola is equal to that which it would acquire in falling through a vertical distance equal to the sum of the sublimity and the altitude of the semi-parabola.*

Let *ab* be a semi-parabola having a sublimity *da* and an altitude *ac*, the sum of which is the perpendicular *dc*. Now I say the momentum of the particle at *b* is the same as that which it would acquire in falling freely from *d* to *c*. Let us take the length of *dc* itself as a measure of time and momentum, and lay off *cf* equal to the mean proportional between *cd* and *da;* also lay off *ce* a mean proportional between *cd* and *ca*. Now *cf* is the measure of the time and of the momentum acquired by fall, from rest at *d*, through the distance *da;* while *ce* is the time and momentum of fall, from rest at *a*, through the distance *ca;* also the diagonal *ef* will represent a momentum which is the resultant of these two, and is therefore the momentum at the terminal point of the parabola, *b*.

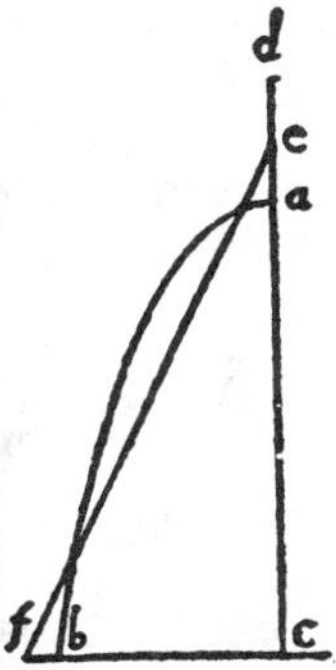

Fig. 122

And since *dc* has been cut at some point *a* and since *cf* and *ce* are mean proportionals between the whole of *cd* and its parts, *da* and *ac*, it follows, from the preceding lemma, that the sum of the squares of these mean proportionals is equal to the square of the whole: but the square of *ef* is also equal to the sum of these same squares; whence it follows that the line *ef* is equal to *dc*.

Accordingly the momentum acquired at *c* by a particle in falling from *d* is the same as that acquired at *b* by a particle traversing the parabola *ab*.   Q. E. D.

* In modern mechanics this well-known theorem assumes the following form: *The speed of a projectile at any point is that produced by a fall from the directrix.* [*Trans.*]

COROLLARY

Hence it follows that, in the case of all parabolas where the sum of the sublimity and altitude is a constant, the momentum at the terminal point is a constant.

PROBLEM. PROPOSITION XI

Given the amplitude and the speed [*impetus*] at the terminal point of a semi-parabola, to find its altitude.

Let the given speed be represented by the vertical line *ab*, and the amplitude by the horizontal line *bc*; it is required to find the sublimity of the semi-parabola whose terminal speed is *ab* and amplitude *bc*. From what precedes [Cor. Prop. V] it is clear that half the amplitude *bc* is a mean proportional between

[300]

the altitude and sublimity of the parabola of which the terminal speed is equal, in accordance with the preceding proposition, to the speed acquired by a body in falling from rest at *a* through the distance *ab*. Therefore the line *ba* must be cut at a point such that the rectangle formed by its two parts will be equal to the square of half *bc*, namely *bd*. Necessarily, therefore, *bd* must not exceed the half of *ba*; for of all the rectangles formed by parts of a straight line the one of greatest area is obtained when the line is divided into two equal parts. Let *e* be the middle point of the line *ab*; and now if *bd* be equal to *be* the problem is solved; for *be* will be the altitude and *ea* the sublimity of the parabola. (Incidentally we may observe a consequence already demonstrated, namely: of all parabolas described with any given terminal speed that for which the elevation is 45° will have the maximum amplitude.)

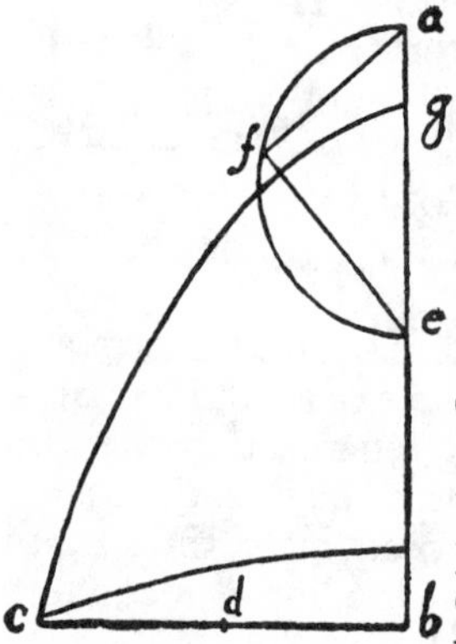

Fig. 123

But suppose that *bd* is less than half of *ba* which is to be divided

divided in such a way that the rectangle upon its parts may be equal to the square of *bd*. Upon *ea* as diameter describe a semi-circle *efa*, in which draw the chord *af*, equal to *bd:* join *fe* and lay off the distance *eg* equal to *fe*. Then the rectangle *bg.ga* plus the square of *eg* will be equal to the square of *ea*, and hence also to the sum of the squares of *af* and *fe*. If now we subtract the equal squares of *fe* and *ge* there remains the rectangle *bg.ga* equal to the square of *af*, that is, of *bd*, a line which is a mean proportional between *bg* and *ga;* from which it is evident that the semi-parabola whose amplitude is *bc* and whose terminal speed [*impetus*] is represented by *ba* has an altitude *bg* and a sublimity *ga*.

If however we lay off *bi* equal to *ga*, then *bi* will be the altitude of the semi-parabola *ic*, and *ia* will be its sublimity. From the preceding demonstration we are able to solve the following problem.

PROBLEM. PROPOSITION XII

To compute and tabulate the amplitudes of all semi-parabolas which are described by projectiles fired with the same initial speed [*impetus*].

From the foregoing it follows that, whenever the sum of the altitude and sublimity is a constant vertical height for any set of parabolas, these parabolas are described by projectiles having the same initial speed; all vertical heights thus

[301]

obtained are therefore included between two parallel horizontal lines. Let *cb* represent a horizontal line and *ab* a vertical line of equal length; draw the diagonal *ac;* the angle *acb* will be one of 45°; let *d* be the middle point of the vertical line *ab*. Then the semi-parabola *dc* is the one which is determined by the sublimity *ad* and the altitude *db*, while its terminal speed at *c* is that which would be acquired at *b* by a particle falling from rest at *a*. If now *ag* be drawn parallel to *bc*, the sum of the altitude and sublimity for any other semi-parabola having the same terminal speed will, in the manner explained, be equal to the distance between the parallel lines *ag* and *bc*. Moreover, since

it

it has already been shown that the amplitudes of two semi-parabolas are the same when their angles of elevation differ from 45° by like amounts, it follows that the same computation which is employed for the larger elevation will serve also for the smaller.

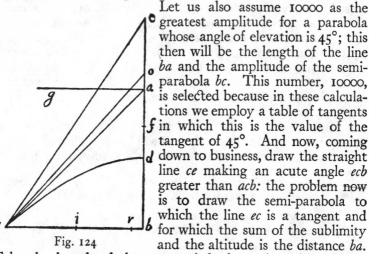

Let us also assume 10000 as the greatest amplitude for a parabola whose angle of elevation is 45°; this then will be the length of the line *ba* and the amplitude of the semi-parabola *bc*. This number, 10000, is selected because in these calculations we employ a table of tangents in which this is the value of the tangent of 45°. And now, coming down to business, draw the straight line *ce* making an acute angle *ecb* greater than *acb*: the problem now is to draw the semi-parabola to which the line *ec* is a tangent and for which the sum of the sublimity and the altitude is the distance *ba*.

Fig. 124

Take the length of the tangent * *be* from the table of tangents, using the angle *bce* as an argument: let *f* be the middle point of *be;* next find a third proportional to *bf* and *bi* (the half of *bc*), which is of necessity greater than *fa*.† Call this *fo*. We have now discovered that, for the parabola inscribed

[302]

in the triangle *ecb* having the tangent *ce* and the amplitude *cb*, the altitude is *bf* and the sublimity *fo*. But the total length of *bo* exceeds the distance between the parallels *ag* and *cb*, while our problem was to keep it equal to this distance: for both the parabola sought and the parabola *dc* are described

* The reader will observe that the word "tangent" is here used in a sense somewhat different from that of the preceding sentence. The "tangent *ec*" is a line which touches the parabola at *c;* but the "tangent *eb*" is the side of the right-angled triangle which lies opposite the angle *ecb*, a line whose length is proportional to the numerical value of the tangent of this angle. [*Trans.*]

† This fact is demonstrated in the third paragraph below. [*Trans.*]

by projectiles fired from *c* with the same speed. Now since an infinite number of greater and smaller parabolas, similar to each other, may be described within the angle *bce* we must find another parabola which like *cd* has for the sum of its altitude and sublimity the height *ba*, equal to *bc*.

Therefore lay off *cr* so that, *ob:ba = bc:cr;* then *cr* will be the amplitude of a semi-parabola for which *bce* is the angle of elevation and for which the sum of the altitude and sublimity is the distance between the parallels *ga* and *cb*, as desired. The process is therefore as follows: One draws the tangent of the given angle *bce;* takes half of this tangent, and adds to it the quantity, *fo,* which is a third proportional to the half of this tangent and the half of *bc;* the desired amplitude *cr* is then found from the following proportion *ob:ba = bc:cr.* For example let the angle *ecb* be one of 50°; its tangent is 11918, half of which, namely *bf,* is 5959; half of *bc* is 5000; the third proportional of these halves is 4195, which added to *bf* gives the value 10154 for *bo.* Further, as *ob* is to *ab,* that is, as 10154 is to 10000, so is *bc,* or 10000 (each being the tangent of 45°) to *cr,* which is the amplitude sought and which has the value 9848, the maximum amplitude being *bc,* or 10000. The amplitudes of the entire parabolas are double these, namely, 19696 and 20000. This is also the amplitude of a parabola whose angle of elevation is 40°, since it deviates by an equal amount from one of 45°.

[303]

SAGR. In order to thoroughly understand this demonstration I need to be shown how the third proportional of *bf* and *bi* is, as the Author indicates, necessarily greater than *fa.*

SALV. This result can, I think, be obtained as follows. The square of the mean proportional between two lines is equal to the rectangle formed by these two lines. Therefore the square of *bi* (or of *bd* which is equal to *bi*) must be equal to the rectangle formed by *fb* and the desired third proportional. This third proportional is necessarily greater than *fa* because the rectangle formed by *bf* and *fa* is less than the square of *bd* by an amount equal to the square of *df,* as shown in Euclid, II. 1. Besides it is to be observed that the point *f,* which is the middle point of the tangent

tangent *eb*, falls in general above *a* and only once at *a;* in which cases it is self-evident that the third proportional to the half of the tangent and to the sublimity *bi* lies wholly above *a*. But the Author has taken a case where it is not evident that the third proportional is always greater than *fa*, so that when laid off above the point *f* it extends beyond the parallel *ag*.

Now let us proceed. It will be worth while, by the use of this table, to compute another giving the altitudes of these semi-parabolas described by projectiles having the same initial speed. The construction is as follows:

[304]

Amplitudes of semi-parabolas described with the same initial speed.		Altitudes of semi-parabolas described with the same initial speed.	
Angle of Elevation	Angle of Elevation	Angle of Elevation	Angle of Elevation
45° 10000		1° 3	46° 5173
46 9994	44°	2 13	47 5346
47 9976	43	3 28	48 5523
48 9945	42	4 50	49 5698
49 9902	41	5 76	50 5868
50 9848	40	6 108	51 6038
51 9782	39	7 150	52 6207
52 9704	38	8 194	53 6379
53 9612	37	9 245	54 6546
54 9511	36	10 302	55 6710
55 9396	35	11 365	56 6873
56 9272	34	12 432	57 7033
57 9136	33	13 506	58 7190
58 8989	32	14 585	59 7348
59 8829	31	15 670	60 7502
60 8659	30	16 760	61 7649
61 8481	29	17 855	62 7796
62 8290	28	18 955	63 7939
63 8090	27	19 1060	64 8078
64 7880	26	20 1170	65 8214
65 7660	25	21 1285	66 8346
66 7431	24	22 1402	67 8474
67 7191	23	23 1527	68 8597
68 6944	22	24 1685	69 8715
69 6692	21	25 1786	70 8830

Amplitudes of semi-parabolas described with the same initial speed.

Angle of Elevation		Angle of Elevation
70°	6428	20°
71	6157	19
72	5878	18
73	5592	17
74	5300	16
75	5000	15
76	4694	14
77	4383	13
78	4067	12
79	3746	11
80	3420	10
81	3090	9
82	2756	8
83	2419	7
84	2079	6
85	1736	5
86	1391	4
87	1044	3
88	698	2
89	349	1

Altitudes of semi-parabolas described with the same initial speed.

Angle of Elevation		Angle of Elevation	
26°	1922	71°	8940
27	2061	72	9045
28	2204	73	9144
29	2351	74	9240
30	2499	75	9330
31	2653	76	9415
32	2810	77	9493
33	2967	78	9567
34	3128	79	9636
35	3289	80	9698
36	3456	81	9755
37	3621	82	9806
38	3793	83	9851
39	3962	84	9890
40	4132	85	9924
41	4302	86	9951
42	4477	87	9972
43	4654	88	9987
44	4827	89	9998
45	5000	90	10000

[305]

PROBLEM. PROPOSITION XIII

From the amplitudes of semi-parabolas given in the preceding table to find the altitudes of each of the parabolas described with the same initial speed.

Let bc denote the given amplitude; and let ob, the sum of the altitude and sublimity, be the measure of the initial speed which is understood to remain constant. Next we must find and determine the altitude, which we shall accomplish by so dividing ob that the rectangle contained by its parts shall be equal to the square of half the amplitude, bc. Let f denote this point of division and d and i be the middle points of ob and bc respectively. The square of ib is equal to the rectangle $bf.fo$; but the square of do is equal to the sum of the rectangle $bf.fo$ and the

square

square of *fd*. If, therefore, from the square of *do* we subtract the square of *bi* which is equal to the rectangle *bf.fo*, there will remain the square of *fd*. The altitude in question, *bf*, is now obtained by adding to this length, *fd*, the line *bd*. The process is then as follows: From the square of half of *bo* which is known, subtract the square of *bi* which is also known; take the square root of the remainder and add to it the known length *db;* then you have the required altitude, *bf*.

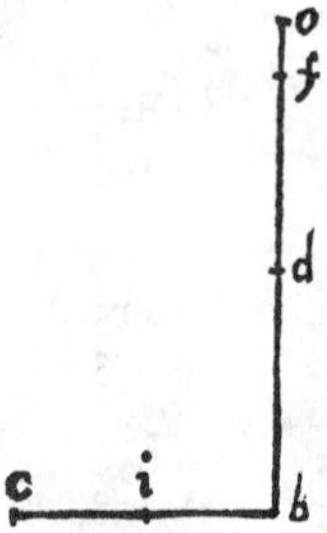

Example. To find the altitude of a semi-parabola described with an angle of elevation of 55°. From the preceding table the amplitude is seen to be 9396, of which the half is 4698, and the square 22071204. When this is subtracted from the square of the half of *bo*, which is always 25,000,000, the remainder is 2928796, of which the square root

Fig. 125

[306]

is approximately 1710. Adding this to the half of *bo*, namely 5000, we have 6710 for the altitude of *bf*.

It will be worth while to add a third table giving the altitudes and sublimities for parabolas in which the amplitude is a constant.

SAGR. I shall be very glad to see this; for from it I shall learn the difference of speed and force [*degl' impeti e delle forze*] required to fire projectiles over the same range with what we call mortar shots. This difference will, I believe, vary greatly with the elevation so that if, for example, one wished to employ an elevation of 3° or 4°, or 87° or 88° and yet give the ball the same range which it had with an elevation of 45° (where we have shown the initial speed to be a minimum) the excess of force required will, I think, be very great.

SALV. You are quite right, sir; and you will find that in order to perform this operation completely, at all angles of elevation, you will have to make great strides toward an infinite speed. We pass now to the consideration of the table.

[307]

Table giving the altitudes and sublimities of parabolas of constant amplitude, namely 10000, computed for each degree of elevation.

Angle of Elevation	Altitude	Sublimity	Angle of Elevation	Altitude	Sublimity
1°	87	286533	46°	5177	4828
2	175	142450	47	5363	4662
3	262	95802	48	5553	4502
4	349	71531	49	5752	4345
5	437	57142	50	5959	4196
6	525	47573	51	6174	4048
7	614	40716	52	6399	3906
8	702	35587	53	6635	3765
9	792	31565	54	6882	3632
10	881	28367	55	7141	3500
11	972	25720	56	7413	3372
12	1063	23518	57	7699	3247
13	1154	21701	58	8002	3123
14	1246	20056	59	8332	3004
15	1339	18663	60	8600	2887
16	1434	17405	61	9020	2771
17	1529	16355	62	9403	2658
18	1624	15389	63	9813	2547
19	1722	14522	64	10251	2438
20	1820	13736	65	10722	2331
21	1919	13024	66	11230	2226
22	2020	12376	67	11779	2122
23	2123	11778	68	12375	2020
24	2226	11230	69	13025	1919
25	2332	10722	70	13237	1819
26	2439	10253	71	14521	1721
27	2547	9814	72	15388	1624
28	2658	9404	73	16354	1528
29	2772	9020	74	17437	1433
30	2887	8659	75	18660	1339
31	3008	8336	76	20054	1246
32	3124	8001	77	21657	1154
33	3247	7699	78	23523	1062
34	3373	7413	79	25723	972
35	3501	7141	80	28356	881
36	3633	6882	81	31569	792
37	3768	6635	82	35577	702
38	3906	6395	83	40222	613

Angle of Elevation	Altitude	Sublimity	Angle of Elevation	Altitude	Sublimity
39°	4049	6174	84°	47572	525
40	4196	5959	85	57150	437
41	4346	5752	86	71503	349
42	4502	5553	87	95405	262
43	4662	5362	88	143181	174
44	4828	5177	89	286499	87
45	5000	5000	90	infinita	

[308]
PROPOSITION XIV

To find for each degree of elevation the altitudes and sublimities of parabolas of constant amplitude.

The problem is easily solved. For if we assume a constant amplitude of 10000, then half the tangent at any angle of elevation will be the altitude. Thus, to illustrate, a parabola having an angle of elevation of 30° and an amplitude of 10000, will have an altitude of 2887, which is approximately one-half the tangent. And now the altitude having been found, the sublimity is derived as follows. Since it has been proved that half the amplitude of a semi-parabola is the mean proportional between the altitude and sublimity, and since the altitude has already been found, and since the semi-amplitude is a constant, namely 5000, it follows that if we divide the square of the semi-amplitude by the altitude we shall obtain the sublimity sought. Thus in our example the altitude was found to be 2887: the square of 5000 is 25,000,000, which divided by 2887 gives the approximate value of the sublimity, namely 8659.

SALV. Here we see, first of all, how very true is the statement made above, that, for different angles of elevation, the greater the deviation from the mean, whether above or below, the greater the initial speed [*impeto e violenza*] required to carry the projectile over the same range. For since the speed is the resultant of two motions, namely, one horizontal and uniform, the other vertical and naturally accelerated; and since the sum of the altitude and sublimity represents this speed, it is seen from the preceding table that this sum is a minimum

minimum for an elevation of 45° where the altitude and sublimity are equal, namely, each 5000; and their sum 10000.
But if we choose a greater elevation, say 50°, we shall find the altitude 5959, and the sublimity 4196, giving a sum of 10155; in like
manner we shall find that this is precisely the value of the speed
at 40° elevation, both angles deviating equally from the mean.

Secondly it is to be noted that, while equal speeds are required for each of two elevations that are equidistant from the
mean, there is this curious alternation, namely, that the altitude
and sublimity at the greater elevation correspond inversely to
the sublimity and altitude at the lower elevation. Thus in the
[309]
preceding example an elevation of 50° gives an altitude of 5959
and a sublimity of 4196; while an elevation of 40° corresponds
to an altitude of 4196 and a sublimity of 5959. And this holds
true in general; but it is to be remembered that, in order to
escape tedious calculations, no account has been taken of
fractions which are of little moment in comparison with such
large numbers.

SAGR. I note also in regard to the two components of the
initial speed [impeto] that the higher the shot the less is the
horizontal and the greater the vertical component; on the other
hand, at lower elevations where the shot reaches only a small
height the horizontal component of the initial speed must be
great. In the case of a projectile fired at an elevation of 90°,
I quite understand that all the force [forza] in the world would
not be sufficient to make it deviate a single finger's breadth from
the perpendicular and that it would necessarily fall back into
its initial position; but in the case of zero elevation, when the
shot is fired horizontally, I am not so certain that some force,
less than infinite, would not carry the projectile some distance;
thus not even a cannon can fire a shot in a perfectly horizontal
direction, or as we say, point blank, that is, with no elevation at
all. Here I admit there is some room for doubt. The fact I do
not deny outright, because of another phenomenon apparently
no less remarkable, but yet one for which I have conclusive
evidence. This phenomenon is the impossibility of stretching

a

a rope in such a way that it shall be at once straight and parallel to the horizon; the fact is that the cord always sags and bends and that no force is sufficient to stretch it perfectly straight.

SALV. In this case of the rope then, Sagredo, you cease to wonder at the phenomenon because you have its demonstration; but if we consider it with more care we may possibly discover some correspondence between the case of the gun and that of the string. The curvature of the path of the shot fired horizontally appears to result from two forces, one (that of the weapon) drives it horizontally and the other (its own weight) draws it vertically downward. So in stretching the rope you have the force which pulls it horizontally and its own weight which acts downwards. The circumstances in these two cases are, therefore, very similar. If then you attribute to the weight of the rope a power and

[310]

energy [*possanza ed energia*] sufficient to oppose and overcome any stretching force, no matter how great, why deny this power to the bullet?

Besides I must tell you something which will both surprise and please you, namely, that a cord stretched more or less tightly assumes a curve which closely approximates the parabola. This similarity is clearly seen if you draw a parabolic curve on a vertical plane and then invert it so that the apex will lie at the bottom and the base remain horizontal; for, on hanging a chain below the base, one end attached to each extremity of the base, you will observe that, on slackening the chain more or less, it bends and fits itself to the parabola; and the coincidence is more exact in proportion as the parabola is drawn with less curvature or, so to speak, more stretched; so that using parabolas described with elevations less than 45° the chain fits its parabola almost perfectly.

SAGR. Then with a fine chain one would be able to quickly draw many parabolic lines upon a plane surface.

SALV. Certainly and with no small advantage as I shall show you later.

SIMP. But before going further, I am anxious to be convinced at least of that proposition of which you say that there is a rigid

rigid demonstration; I refer to the statement that it is impossible by any force whatever to stretch a cord so that it will lie perfectly straight and horizontal.

Sagr. I will see if I can recall the demonstration; but in order to understand it, Simplicio, it will be necessary for you to take for granted concerning machines what is evident not alone from experiment but also from theoretical considerations, namely, that the velocity of a moving body [*velocità del movente*], even when its force [*forza*] is small, can overcome a very great resistance exerted by a slowly moving body, whenever the velocity of the moving body bears to that of the resisting body a greater ratio than the resistance [*resistenza*] of the resisting body to the force [*forza*] of the moving body.

Simp. This I know very well for it has been demonstrated by Aristotle in his *Questions in Mechanics;* it is also clearly seen in the lever and the steelyard where a counterpoise weighing not more than 4 pounds will lift a weight of 400 provided that the distance of the counterpoise from the axis about which the steelyard rotates be more than one hundred times as great as the distance between this axis and the point of support for

[311]

the large weight. This is true because the counterpoise in its descent traverses a space more than one hundred times as great as that moved over by the large weight in the same time; in other words the small counterpoise moves with a velocity which is more than one hundred times as great as that of the large weight.

Sagr. You are quite right; you do not hesitate to admit that however small the force [*forza*] of the moving body it will overcome any resistance, however great, provided it gains more in velocity than it loses in force and weight [*vigore e gravità*]. Now let us return to the case of the cord. In the accompanying figure *ab* represents a line passing through two fixed points *a* and *b;* at the extremities of this line hang, as you see, two large weights *c* and *d,* which stretch it with great force and keep it truly straight, seeing that it is merely a line without weight. Now I wish to remark that if from the middle point of this line, which

which we may call e, you suspend any small weight, say h, the line ab will yield toward the point f and on account of its elongation will compel the two heavy weights c and d to rise. This I shall demonstrate as follows: with the points a and b as centers describe the two quadrants, eig and elm; now since the two semi-diameters ai and bl are equal to ae and eb, the remainders fi and fl are the excesses of the lines af and fb over ae and eb; they there-

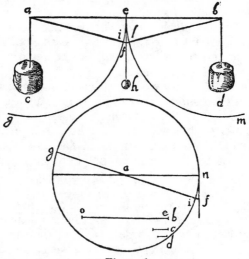

Fig. 126

fore determine the rise of the weights c and d, assuming of course that the weight h has taken the position f. But the weight h

[312]

will take the position f, whenever the line ef which represents the descent of h bears to the line fi—that is, to the rise of the weights c and d—a ratio which is greater than the ratio of the weight of the two large bodies to that of the body h. Even when the weights of c and d are very great and that of h very small this will happen; for the excess of the weights c and d over the weight of h can never be so great but that the excess of the tangent ef over the segment fi may be proportionally greater. This may

be

be proved as follows: Draw a circle of diameter *gai;* draw the line *bo* such that the ratio of its length to another length *c*, *c*>*d*, is the same as the ratio of the weights *c* and *d* to the weight *h*. Since *c*>*d*, the ratio of *bo* to *d* is greater than that of *bo* to *c*. Take *be* a third proportional to *ob* and *d;* prolong the diameter *gi* to a point *f* such that *gi:if* = *oe:eb;* and from the point *f* draw the tangent *fn;* then since we already have *oe:eb* = *gi:if,* we shall obtain, by compounding ratios, *ob:eb* = *gf:if.* But *d* is a mean proportional between *ob* and *be;* while *nf* is a mean proportional between *gf* and *fi.* Hence *nf* bears to *fi* the same ratio as that of *cb* to *d*, which is greater than that of the weights *c* and *d* to the weight *h*. Since then the descent, or velocity, of the weight *h* bears to the rise, or velocity, of the weights *c* and *d* a greater ratio than the weight of the bodies *c* and *d* bears to the weight of *h*, it is clear that the weight *h* will descend and the line *ab* will cease to be straight and horizontal.

And now this which happens in the case of a weightless cord *ab* when any small weight *h* is attached at the point *e*, happens also when the cord is made of ponderable matter but without any attached weight; because in this case the material of which the cord is composed functions as a suspended weight.

SIMP. I am fully satisfied. So now Salviati can explain, as he promised, the advantage of such a chain and, afterwards, present the speculations of our Academician on the subject of impulsive forces [*forza della percossa*].

SALV. Let the preceding discussions suffice for to-day; the hour is already late and the time remaining will not permit us to clear up the subjects proposed; we may therefore postpone our meeting until another and more opportune occasion.

SAGR. I concur in your opinion, because after various conversations with intimate friends of our Academician I have concluded that this question of impulsive forces is very obscure, and I think that, up to the present, none of those who have treated

[313]

this subject have been able to clear up its dark corners which lie almost beyond the reach of human imagination; among the various views which I have heard expressed one, strangely fantastic

tastic, remains in my memory, namely, that impulsive forces are indeterminate, if not infinite. Let us, therefore, await the convenience of Salviati. Meanwhile tell me what is this which follows the discussion of projectiles.

SALV. These are some theorems pertaining to the centers of gravity of solids, discovered by our Academician in his youth, and undertaken by him because he considered the treatment of Federigo Comandino to be somewhat incomplete. The propositions which you have before you would, he thought, meet the deficiencies of Comandino's book. The investigation was undertaken at the instance of the Illustrious Marquis Guid' Ubaldo Dal Monte, a very distinguished mathematician of his day, as is evidenced by his various publications. To this gentleman our Academician gave a copy of this work, hoping to extend the investigation to other solids not treated by Comandino. But a little later there chanced to fall into his hands the book of the great geometrician, Luca Valerio, where he found the subject treated so completely that he left off his own investigations, although the methods which he employed were quite different from those of Valerio.

SAGR. Please be good enough to leave this volume with me until our next meeting so that I may be able to read and study these propositions in the order in which they are written.

SALV. It is a pleasure to comply with your request and I only hope that the propositions will be of deep interest to you.

END OF FOURTH DAY.

APPENDIX

Containing some theorems, and their proofs, dealing with centers of gravity of solid bodies, written by the same Author at an earlier date.*

* Following the example of the National Edition, this *Appendix* which covers 18 pages of the Leyden Edition of 1638 is here omitted as being of minor interest. [*Trans.*]

[FINIS]

INDEX

A CATALOG OF SELECTED
DOVER BOOKS
IN ALL FIELDS OF INTEREST

A CATALOG OF SELECTED DOVER
BOOKS IN ALL FIELDS OF INTEREST

DRAWINGS OF REMBRANDT, edited by Seymour Slive. Updated Lippmann, Hofstede de Groot edition, with definitive scholarly apparatus. All portraits, biblical sketches, landscapes, nudes. Oriental figures, classical studies, together with selection of work by followers. 550 illustrations. Total of 630pp. 9⅛ × 12¼.
21485-0, 21486-9 Pa., Two-vol. set $25.00

GHOST AND HORROR STORIES OF AMBROSE BIERCE, Ambrose Bierce. 24 tales vividly imagined, strangely prophetic, and decades ahead of their time in technical skill: "The Damned Thing," "An Inhabitant of Carcosa," "The Eyes of the Panther," "Moxon's Master," and 20 more. 199pp. 5⅜ × 8½. 20767-6 Pa. $3.95

ETHICAL WRITINGS OF MAIMONIDES, Maimonides. Most significant ethical works of great medieval sage, newly translated for utmost precision, readability. Laws Concerning Character Traits, Eight Chapters, more. 192pp. 5⅜ × 8½.
24522-5 Pa. $4.50

THE EXPLORATION OF THE COLORADO RIVER AND ITS CANYONS, J. W. Powell. Full text of Powell's 1,000-mile expedition down the fabled Colorado in 1869. Superb account of terrain, geology, vegetation, Indians, famine, mutiny, treacherous rapids, mighty canyons, during exploration of last unknown part of continental U.S. 400pp. 5⅜ × 8½. 20094-9 Pa. $6.95

HISTORY OF PHILOSOPHY, Julián Marías. Clearest one-volume history on the market. Every major philosopher and dozens of others, to Existentialism and later. 505pp. 5⅜ × 8½. 21739-6 Pa. $8.50

ALL ABOUT LIGHTNING, Martin A. Uman. Highly readable non-technical survey of nature and causes of lightning, thunderstorms, ball lightning, St. Elmo's Fire, much more. Illustrated. 192pp. 5⅜ × 8½. 25237-X Pa. $5.95

SAILING ALONE AROUND THE WORLD, Captain Joshua Slocum. First man to sail around the world, alone, in small boat. One of great feats of seamanship told in delightful manner. 67 illustrations. 294pp. 5⅜ × 8½. 20326-3 Pa. $4.95

LETTERS AND NOTES ON THE MANNERS, CUSTOMS AND CONDITIONS OF THE NORTH AMERICAN INDIANS, George Catlin. Classic account of life among Plains Indians: ceremonies, hunt, warfare, etc. 312 plates. 572pp. of text. 6⅛ × 9¼. 22118-0, 22119-9 Pa. Two-vol. set $15.90

ALASKA: The Harriman Expedition, 1899, John Burroughs, John Muir, et al. Informative, engrossing accounts of two-month, 9,000-mile expedition. Native peoples, wildlife, forests, geography, salmon industry, glaciers, more. Profusely illustrated. 240 black-and-white line drawings. 124 black-and-white photographs. 3 maps. Index. 576pp. 5⅜ × 8½. 25109-8 Pa. $11.95

THE BOOK OF BEASTS: Being a Translation from a Latin Bestiary of the Twelfth Century, T. H. White. Wonderful catalog real and fanciful beasts: manticore, griffin, phoenix, amphivius, jaculus, many more. White's witty erudite commentary on scientific, historical aspects. Fascinating glimpse of medieval mind. Illustrated. 296pp. 5⅜ × 8¼. (Available in U.S. only) 24609-4 Pa. $5.95

FRANK LLOYD WRIGHT: ARCHITECTURE AND NATURE With 160 Illustrations, Donald Hoffmann. Profusely illustrated study of influence of nature—especially prairie—on Wright's designs for Fallingwater, Robie House, Guggenheim Museum, other masterpieces. 96pp. 9¼ × 10¾. 25098-9 Pa. $7.95

FRANK LLOYD WRIGHT'S FALLINGWATER, Donald Hoffmann. Wright's famous waterfall house: planning and construction of organic idea. History of site, owners, Wright's personal involvement. Photographs of various stages of building. Preface by Edgar Kaufmann, Jr. 100 illustrations. 112pp. 9¼ × 10.
23671-4 Pa. $7.95

YEARS WITH FRANK LLOYD WRIGHT: Apprentice to Genius, Edgar Tafel. Insightful memoir by a former apprentice presents a revealing portrait of Wright the man, the inspired teacher, the greatest American architect. 372 black-and-white illustrations. Preface. Index. vi + 228pp. 8¼ × 11. 24801-1 Pa. $9.95

THE STORY OF KING ARTHUR AND HIS KNIGHTS, Howard Pyle. Enchanting version of King Arthur fable has delighted generations with imaginative narratives of exciting adventures and unforgettable illustrations by the author. 41 illustrations. xviii + 313pp. 6⅛ × 9¼. 21445-1 Pa. $5.95

THE GODS OF THE EGYPTIANS, E. A. Wallis Budge. Thorough coverage of numerous gods of ancient Egypt by foremost Egyptologist. Information on evolution of cults, rites and gods; the cult of Osiris; the Book of the Dead and its rites; the sacred animals and birds; Heaven and Hell; and more. 956pp. 6⅛ × 9¼.
22055-9, 22056-7 Pa., Two-vol. set $21.90

A THEOLOGICO-POLITICAL TREATISE, Benedict Spinoza. Also contains unfinished *Political Treatise*. Great classic on religious liberty, theory of government on common consent. R. Elwes translation. Total of 421pp. 5⅜ × 8½.
20249-6 Pa. $6.95

INCIDENTS OF TRAVEL IN CENTRAL AMERICA, CHIAPAS, AND YUCATAN, John L. Stephens. Almost single-handed discovery of Maya culture; exploration of ruined cities, monuments, temples; customs of Indians. 115 drawings. 892pp. 5⅜ × 8½. 22404-X, 22405-8 Pa., Two-vol. set $15.90

LOS CAPRICHOS, Francisco Goya. 80 plates of wild, grotesque monsters and caricatures. Prado manuscript included. 183pp. 6⅜ × 9⅜. 22384-1 Pa. $4.95

AUTOBIOGRAPHY: The Story of My Experiments with Truth, Mohandas K. Gandhi. Not hagiography, but Gandhi in his own words. Boyhood, legal studies, purification, the growth of the Satyagraha (nonviolent protest) movement. Critical, inspiring work of the man who freed India. 480pp. 5⅜ × 8½. (Available in U.S. only)
24593-4 Pa. $6.95

ILLUSTRATED DICTIONARY OF HISTORIC ARCHITECTURE, edited by Cyril M. Harris. Extraordinary compendium of clear, concise definitions for over 5,000 important architectural terms complemented by over 2,000 line drawings. Covers full spectrum of architecture from ancient ruins to 20th-century Modernism. Preface. 592pp. 7½ × 9⅝. 24444-X Pa. $14.95

THE NIGHT BEFORE CHRISTMAS, Clement Moore. Full text, and woodcuts from original 1848 book. Also critical, historical material. 19 illustrations. 40pp. 4⅝ × 6. 22797-9 Pa. $2.50

THE LESSON OF JAPANESE ARCHITECTURE: 165 Photographs, Jiro Harada. Memorable gallery of 165 photographs taken in the 1930's of exquisite Japanese homes of the well-to-do and historic buildings. 13 line diagrams. 192pp. 8⅜ × 11¼. 24778-3 Pa. $8.95

THE AUTOBIOGRAPHY OF CHARLES DARWIN AND SELECTED LETTERS, edited by Francis Darwin. The fascinating life of eccentric genius composed of an intimate memoir by Darwin (intended for his children); commentary by his son, Francis; hundreds of fragments from notebooks, journals, papers; and letters to and from Lyell, Hooker, Huxley, Wallace and Henslow. xi + 365pp. 5⅜ × 8. 20479-0 Pa. $5.95

WONDERS OF THE SKY: Observing Rainbows, Comets, Eclipses, the Stars and Other Phenomena, Fred Schaaf. Charming, easy-to-read poetic guide to all manner of celestial events visible to the naked eye. Mock suns, glories, Belt of Venus, more. Illustrated. 299pp. 5¼ × 8¼. 24402-4 Pa. $7.95

BURNHAM'S CELESTIAL HANDBOOK, Robert Burnham, Jr. Thorough guide to the stars beyond our solar system. Exhaustive treatment. Alphabetical by constellation: Andromeda to Cetus in Vol. 1; Chamaeleon to Orion in Vol. 2; and Pavo to Vulpecula in Vol. 3. Hundreds of illustrations. Index in Vol. 3. 2,000pp. 6⅛ × 9¼. 23567-X, 23568-8, 23673-0 Pa., Three-vol. set $37.85

STAR NAMES: Their Lore and Meaning, Richard Hinckley Allen. Fascinating history of names various cultures have given to constellations and literary and folkloristic uses that have been made of stars. Indexes to subjects. Arabic and Greek names. Biblical references. Bibliography. 563pp. 5⅜ × 8½. 21079-0 Pa. $7.95

THIRTY YEARS THAT SHOOK PHYSICS: The Story of Quantum Theory, George Gamow. Lucid, accessible introduction to influential theory of energy and matter. Careful explanations of Dirac's anti-particles, Bohr's model of the atom, much more. 12 plates. Numerous drawings. 240pp. 5⅜ × 8½. 24895-X Pa. $4.95

CHINESE DOMESTIC FURNITURE IN PHOTOGRAPHS AND MEASURED DRAWINGS, Gustav Ecke. A rare volume, now affordably priced for antique collectors, furniture buffs and art historians. Detailed review of styles ranging from early Shang to late Ming. Unabridged republication. 161 black-and-white drawings, photos. Total of 224pp. 8⅜ × 11¼. (Available in U.S. only) 25171-3 Pa. $12.95

VINCENT VAN GOGH: A Biography, Julius Meier-Graefe. Dynamic, penetrating study of artist's life, relationship with brother, Theo, painting techniques, travels, more. Readable, engrossing. 160pp. 5⅜ × 8½. (Available in U.S. only) 25253-1 Pa. $3.95

HOW TO WRITE, Gertrude Stein. Gertrude Stein claimed anyone could understand her unconventional writing—here are clues to help. Fascinating improvisations, language experiments, explanations illuminate Stein's craft and the art of writing. Total of 414pp. 4⅜ × 6⅜. 23144-5 Pa. $5.95

ADVENTURES AT SEA IN THE GREAT AGE OF SAIL: Five Firsthand Narratives, edited by Elliot Snow. Rare true accounts of exploration, whaling, shipwreck, fierce natives, trade, shipboard life, more. 33 illustrations. Introduction. 353pp. 5⅜ × 8½. 25177-2 Pa. $7.95

THE HERBAL OR GENERAL HISTORY OF PLANTS, John Gerard. Classic descriptions of about 2,850 plants—with over 2,700 illustrations—includes Latin and English names, physical descriptions, varieties, time and place of growth, more. 2,706 illustrations. xlv + 1,678pp. 8½ × 12¼. 23147-X Cloth. $75.00

DOROTHY AND THE WIZARD IN OZ, L. Frank Baum. Dorothy and the Wizard visit the center of the Earth, where people are vegetables, glass houses grow and Oz characters reappear. Classic sequel to *Wizard of Oz*. 256pp. 5⅜ × 8. 24714-7 Pa. $4.95

SONGS OF EXPERIENCE: Facsimile Reproduction with 26 Plates in Full Color, William Blake. This facsimile of Blake's original "Illuminated Book" reproduces 26 full-color plates from a rare 1826 edition. Includes "The Tyger," "London," "Holy Thursday," and other immortal poems. 26 color plates. Printed text of poems. 48pp. 5¼ × 7. 24636-1 Pa. $3.50

SONGS OF INNOCENCE, William Blake. The first and most popular of Blake's famous "Illuminated Books," in a facsimile edition reproducing all 31 brightly colored plates. Additional printed text of each poem. 64pp. 5¼ × 7. 22764-2 Pa. $3.50

PRECIOUS STONES, Max Bauer. Classic, thorough study of diamonds, rubies, emeralds, garnets, etc.: physical character, occurrence, properties, use, similar topics. 20 plates, 8 in color. 94 figures. 659pp. 6⅛ × 9¼. 21910-0, 21911-9 Pa., Two-vol. set $15.90

ENCYCLOPEDIA OF VICTORIAN NEEDLEWORK, S. F. A. Caulfeild and Blanche Saward. Full, precise descriptions of stitches, techniques for dozens of needlecrafts—most exhaustive reference of its kind. Over 800 figures. Total of 679pp. 8⅛ × 11. Two volumes. Vol. 1 22800-2 Pa. $11.95
Vol. 2 22801-0 Pa. $11.95

THE MARVELOUS LAND OF OZ, L. Frank Baum. Second Oz book, the Scarecrow and Tin Woodman are back with hero named Tip, Oz magic. 136 illustrations. 287pp. 5⅜ × 8½. 20692-0 Pa. $5.95

WILD FOWL DECOYS, Joel Barber. Basic book on the subject, by foremost authority and collector. Reveals history of decoy making and rigging, place in American culture, different kinds of decoys, how to make them, and how to use them. 140 plates. 156pp. 7⅞ × 10¾. 20011-6 Pa. $8.95

HISTORY OF LACE, Mrs. Bury Palliser. Definitive, profusely illustrated chronicle of lace from earliest times to late 19th century. Laces of Italy, Greece, England, France, Belgium, etc. Landmark of needlework scholarship. 266 illustrations. 672pp. 6⅛ × 9¼. 24742-2 Pa. $14.95

ILLUSTRATED GUIDE TO SHAKER FURNITURE, Robert Meader. All furniture and appurtenances, with much on unknown local styles. 235 photos. 146pp. 9 × 12. 22819-3 Pa. $7.95

WHALE SHIPS AND WHALING: A Pictorial Survey, George Francis Dow. Over 200 vintage engravings, drawings, photographs of barks, brigs, cutters, other vessels. Also harpoons, lances, whaling guns, many other artifacts. Comprehensive text by foremost authority. 207 black-and-white illustrations. 288pp. 6 × 9. 24808-9 Pa. $8.95

THE BERTRAMS, Anthony Trollope. Powerful portrayal of blind self-will and thwarted ambition includes one of Trollope's most heartrending love stories. 497pp. 5⅜ × 8½. 25119-5 Pa. $8.95

ADVENTURES WITH A HAND LENS, Richard Headstrom. Clearly written guide to observing and studying flowers and grasses, fish scales, moth and insect wings, egg cases, buds, feathers, seeds, leaf scars, moss, molds, ferns, common crystals, etc.—all with an ordinary, inexpensive magnifying glass. 209 exact line drawings aid in your discoveries. 220pp. 5⅜ × 8½. 23330-8 Pa. $4.50

RODIN ON ART AND ARTISTS, Auguste Rodin. Great sculptor's candid, wide-ranging comments on meaning of art; great artists; relation of sculpture to poetry, painting, music; philosophy of life, more. 76 superb black-and-white illustrations of Rodin's sculpture, drawings and prints. 119pp. 8⅜ × 11¼. 24487-3 Pa. $6.95

FIFTY CLASSIC FRENCH FILMS, 1912–1982: A Pictorial Record, Anthony Slide. Memorable stills from Grand Illusion, Beauty and the Beast, Hiroshima, Mon Amour, many more. Credits, plot synopses, reviews, etc. 160pp. 8¼ × 11. 25256-6 Pa. $11.95

THE PRINCIPLES OF PSYCHOLOGY, William James. Famous long course complete, unabridged. Stream of thought, time perception, memory, experimental methods; great work decades ahead of its time. 94 figures. 1,391pp. 5⅜ × 8½. 20381-6, 20382-4 Pa., Two-vol. set $19.90

BODIES IN A BOOKSHOP, R. T. Campbell. Challenging mystery of blackmail and murder with ingenious plot and superbly drawn characters. In the best tradition of British suspense fiction. 192pp. 5⅜ × 8½. 24720-1 Pa. $3.95

CALLAS: PORTRAIT OF A PRIMA DONNA, George Jellinek. Renowned commentator on the musical scene chronicles incredible career and life of the most controversial, fascinating, influential operatic personality of our time. 64 black-and-white photographs. 416pp. 5⅜ × 8¼. 25047-4 Pa. $7.95

GEOMETRY, RELATIVITY AND THE FOURTH DIMENSION, Rudolph Rucker. Exposition of fourth dimension, concepts of relativity as Flatland characters continue adventures. Popular, easily followed yet accurate, profound. 141 illustrations. 133pp. 5⅜ × 8½. 23400-2 Pa. $3.50

HOUSEHOLD STORIES BY THE BROTHERS GRIMM, with pictures by Walter Crane. 53 classic stories—Rumpelstiltskin, Rapunzel, Hansel and Gretel, the Fisherman and his Wife, Snow White, Tom Thumb, Sleeping Beauty, Cinderella, and so much more—lavishly illustrated with original 19th century drawings. 114 illustrations. x + 269pp. 5⅜ × 8½. 21080-4 Pa. $4.50

SUNDIALS, Albert Waugh. Far and away the best, most thorough coverage of ideas, mathematics concerned, types, construction, adjusting anywhere. Over 100 illustrations. 230pp. 5⅜ × 8½. 22947-5 Pa. $4.50

PICTURE HISTORY OF THE NORMANDIE: With 190 Illustrations, Frank O. Braynard. Full story of legendary French ocean liner: Art Deco interiors, design innovations, furnishings, celebrities, maiden voyage, tragic fire, much more. Extensive text. 144pp. 8⅜ × 11¾. 25257-4 Pa. $9.95

THE FIRST AMERICAN COOKBOOK: A Facsimile of "American Cookery," 1796, Amelia Simmons. Facsimile of the first American-written cookbook published in the United States contains authentic recipes for colonial favorites—pumpkin pudding, winter squash pudding, spruce beer, Indian slapjacks, and more. Introductory Essay and Glossary of colonial cooking terms. 80pp. 5⅜ × 8½. 24710-4 Pa. $3.50

101 PUZZLES IN THOUGHT AND LOGIC, C. R. Wylie, Jr. Solve murders and robberies, find out which fishermen are liars, how a blind man could possibly identify a color—purely by your own reasoning! 107pp. 5⅜ × 8½. 20367-0 Pa. $2.50

THE BOOK OF WORLD-FAMOUS MUSIC—CLASSICAL, POPULAR AND FOLK, James J. Fuld. Revised and enlarged republication of landmark work in musico-bibliography. Full information about nearly 1,000 songs and compositions including first lines of music and lyrics. New supplement. Index. 800pp. 5⅜ × 8¼. 24857-7 Pa. $14.95

ANTHROPOLOGY AND MODERN LIFE, Franz Boas. Great anthropologist's classic treatise on race and culture. Introduction by Ruth Bunzel. Only inexpensive paperback edition. 255pp. 5⅜ × 8½. 25245-0 Pa. $5.95

THE TALE OF PETER RABBIT, Beatrix Potter. The inimitable Peter's terrifying adventure in Mr. McGregor's garden, with all 27 wonderful, full-color Potter illustrations. 55pp. 4¼ × 5½. (Available in U.S. only) 22827-4 Pa. $1.75

THREE PROPHETIC SCIENCE FICTION NOVELS, H. G. Wells. *When the Sleeper Wakes, A Story of the Days to Come* and *The Time Machine* (full version). 335pp. 5⅜ × 8½. (Available in U.S. only) 20605-X Pa. $5.95

APICIUS COOKERY AND DINING IN IMPERIAL ROME, edited and translated by Joseph Dommers Vehling. Oldest known cookbook in existence offers readers a clear picture of what foods Romans ate, how they prepared them, etc. 49 illustrations. 301pp. 6⅛ × 9¼. 23563-7 Pa. $6.50

SHAKESPEARE LEXICON AND QUOTATION DICTIONARY, Alexander Schmidt. Full definitions, locations, shades of meaning of every word in plays and poems. More than 50,000 exact quotations. 1,485pp. 6½ × 9¼. 22726-X, 22727-8 Pa., Two-vol. set $27.90

THE WORLD'S GREAT SPEECHES, edited by Lewis Copeland and Lawrence W. Lamm. Vast collection of 278 speeches from Greeks to 1970. Powerful and effective models; unique look at history. 842pp. 5⅜ × 8½. 20468-5 Pa. $11.95

THE BLUE FAIRY BOOK, Andrew Lang. The first, most famous collection, with many familiar tales: Little Red Riding Hood, Aladdin and the Wonderful Lamp, Puss in Boots, Sleeping Beauty, Hansel and Gretel, Rumpelstiltskin; 37 in all. 138 illustrations. 390pp. 5⅜ × 8½.   21437-0 Pa. $5.95

THE STORY OF THE CHAMPIONS OF THE ROUND TABLE, Howard Pyle. Sir Launcelot, Sir Tristram and Sir Percival in spirited adventures of love and triumph retold in Pyle's inimitable style. 50 drawings, 31 full-page. xviii + 329pp. 6½ × 9¼.   21883-X Pa. $6.95

AUDUBON AND HIS JOURNALS, Maria Audubon. Unmatched two-volume portrait of the great artist, naturalist and author contains his journals, an excellent biography by his granddaughter, expert annotations by the noted ornithologist, Dr. Elliott Coues, and 37 superb illustrations. Total of 1,200pp. 5⅜ × 8.
Vol. I 25143-8 Pa. $8.95
Vol. II 25144-6 Pa. $8.95

GREAT DINOSAUR HUNTERS AND THEIR DISCOVERIES, Edwin H. Colbert. Fascinating, lavishly illustrated chronicle of dinosaur research, 1820's to 1960. Achievements of Cope, Marsh, Brown, Buckland, Mantell, Huxley, many others. 384pp. 5¼ × 8¼.   24701-5 Pa. $6.95

THE TASTEMAKERS, Russell Lynes. Informal, illustrated social history of American taste 1850's–1950's. First popularized categories Highbrow, Lowbrow, Middlebrow. 129 illustrations. New (1979) afterword. 384pp. 6 × 9.
23993-4 Pa. $6.95

DOUBLE CROSS PURPOSES, Ronald A. Knox. A treasure hunt in the Scottish Highlands, an old map, unidentified corpse, surprise discoveries keep reader guessing in this cleverly intricate tale of financial skullduggery. 2 black-and-white maps. 320pp. 5⅜ × 8½. (Available in U.S. only)   25032-6 Pa. $5.95

AUTHENTIC VICTORIAN DECORATION AND ORNAMENTATION IN FULL COLOR: 46 Plates from "Studies in Design," Christopher Dresser. Superb full-color lithographs reproduced from rare original portfolio of a major Victorian designer. 48pp. 9¼ × 12¼.   25083-0 Pa. $7.95

PRIMITIVE ART, Franz Boas. Remains the best text ever prepared on subject, thoroughly discussing Indian, African, Asian, Australian, and, especially, Northern American primitive art. Over 950 illustrations show ceramics, masks, totem poles, weapons, textiles, paintings, much more. 376pp. 5⅜ × 8. 20025-6 Pa. $6.95

SIDELIGHTS ON RELATIVITY, Albert Einstein. Unabridged republication of two lectures delivered by the great physicist in 1920–21. *Ether and Relativity* and *Geometry and Experience*. Elegant ideas in non-mathematical form, accessible to intelligent layman. vi + 56pp. 5⅜ × 8½.   24511-X Pa. $2.95

THE WIT AND HUMOR OF OSCAR WILDE, edited by Alvin Redman. More than 1,000 ripostes, paradoxes, wisecracks: Work is the curse of the drinking classes, I can resist everything except temptation, etc. 258pp. 5⅜ × 8½.   20602-5 Pa. $4.50

ADVENTURES WITH A MICROSCOPE, Richard Headstrom. 59 adventures with clothing fibers, protozoa, ferns and lichens, roots and leaves, much more. 142 illustrations. 232pp. 5⅜ × 8½.   23471-1 Pa. $3.95

PLANTS OF THE BIBLE, Harold N. Moldenke and Alma L. Moldenke. Standard reference to all 230 plants mentioned in Scriptures. Latin name, biblical reference, uses, modern identity, much more. Unsurpassed encyclopedic resource for scholars, botanists, nature lovers, students of Bible. Bibliography. Indexes. 123 black-and-white illustrations. 384pp. 6 × 9. 25069-5 Pa. $8.95

FAMOUS AMERICAN WOMEN: A Biographical Dictionary from Colonial Times to the Present, Robert McHenry, ed. From Pocahontas to Rosa Parks, 1,035 distinguished American women documented in separate biographical entries. Accurate, up-to-date data, numerous categories, spans 400 years. Indices. 493pp. 6½ × 9¼. 24523-3 Pa. $9.95

THE FABULOUS INTERIORS OF THE GREAT OCEAN LINERS IN HISTORIC PHOTOGRAPHS, William H. Miller, Jr. Some 200 superb photographs capture exquisite interiors of world's great "floating palaces"—1890's to 1980's: *Titanic, Ile de France, Queen Elizabeth, United States, Europa,* more. Approx. 200 black-and-white photographs. Captions. Text. Introduction. 160pp. 8⅞ × 11¾. 24756-2 Pa. $9.95

THE GREAT LUXURY LINERS, 1927–1954: A Photographic Record, William H. Miller, Jr. Nostalgic tribute to heyday of ocean liners. 186 photos of Ile de France, Normandie, Leviathan, Queen Elizabeth, United States, many others. Interior and exterior views. Introduction. Captions. 160pp. 9 × 12. 24056-8 Pa. $9.95

A NATURAL HISTORY OF THE DUCKS, John Charles Phillips. Great landmark of ornithology offers complete detailed coverage of nearly 200 species and subspecies of ducks: gadwall, sheldrake, merganser, pintail, many more. 74 full-color plates, 102 black-and-white. Bibliography. Total of 1,920pp. 8⅝ × 11¼. 25141-1, 25142-X Cloth. Two-vol. set $100.00

THE SEAWEED HANDBOOK: An Illustrated Guide to Seaweeds from North Carolina to Canada, Thomas F. Lee. Concise reference covers 78 species. Scientific and common names, habitat, distribution, more. Finding keys for easy identification. 224pp. 5⅜ × 8½. 25215-9 Pa. $5.95

THE TEN BOOKS OF ARCHITECTURE: The 1755 Leoni Edition, Leon Battista Alberti. Rare classic helped introduce the glories of ancient architecture to the Renaissance. 68 black-and-white plates. 336pp. 8⅜ × 11¼. 25239-6 Pa. $14.95

MISS MACKENZIE, Anthony Trollope. Minor masterpieces by Victorian master unmasks many truths about life in 19th-century England. First inexpensive edition in years. 392pp. 5⅜ × 8½. 25201-9 Pa. $7.95

THE RIME OF THE ANCIENT MARINER, Gustave Doré, Samuel Taylor Coleridge. Dramatic engravings considered by many to be his greatest work. The terrifying space of the open sea, the storms and whirlpools of an unknown ocean, the ice of Antarctica, more—all rendered in a powerful, chilling manner. Full text. 38 plates. 77pp. 9¼ × 12. 22305-1 Pa. $4.95

THE EXPEDITIONS OF ZEBULON MONTGOMERY PIKE, Zebulon Montgomery Pike. Fascinating first-hand accounts (1805–6) of exploration of Mississippi River, Indian wars, capture by Spanish dragoons, much more. 1,088pp. 5⅜ × 8½. 25254-X, 25255-8 Pa. Two-vol. set $23.90

A CONCISE HISTORY OF PHOTOGRAPHY: Third Revised Edition, Helmut Gernsheim. Best one-volume history—camera obscura, photochemistry, daguerreotypes, evolution of cameras, film, more. Also artistic aspects—landscape, portraits, fine art, etc. 281 black-and-white photographs. 26 in color. 176pp. 8⅜ × 11¼. 25128-4 Pa. $12.95

THE DORÉ BIBLE ILLUSTRATIONS, Gustave Doré. 241 detailed plates from the Bible: the Creation scenes, Adam and Eve, Flood, Babylon, battle sequences, life of Jesus, etc. Each plate is accompanied by the verses from the King James version of the Bible. 241pp. 9 × 12. 23004-X Pa. $8.95

HUGGER-MUGGER IN THE LOUVRE, Elliot Paul. Second Homer Evans mystery-comedy. Theft at the Louvre involves sleuth in hilarious, madcap caper. "A knockout."—Books. 336pp. 5⅜ × 8½. 25185-3 Pa. $5.95

FLATLAND, E. A. Abbott. Intriguing and enormously popular science-fiction classic explores the complexities of trying to survive as a two-dimensional being in a three-dimensional world. Amusingly illustrated by the author. 16 illustrations. 103pp. 5⅜ × 8½. 20001-9 Pa. $2.25

THE HISTORY OF THE LEWIS AND CLARK EXPEDITION, Meriwether Lewis and William Clark, edited by Elliott Coues. Classic edition of Lewis and Clark's day-by-day journals that later became the basis for U.S. claims to Oregon and the West. Accurate and invaluable geographical, botanical, biological, meteorological and anthropological material. Total of 1,508pp. 5⅜ × 8½.
21268-8, 21269-6, 21270-X Pa. Three-vol. set $25.50

LANGUAGE, TRUTH AND LOGIC, Alfred J. Ayer. Famous, clear introduction to Vienna, Cambridge schools of Logical Positivism. Role of philosophy, elimination of metaphysics, nature of analysis, etc. 160pp. 5⅜ × 8½. (Available in U.S. and Canada only) 20010-8 Pa. $2.95

MATHEMATICS FOR THE NONMATHEMATICIAN, Morris Kline. Detailed, college-level treatment of mathematics in cultural and historical context, with numerous exercises. For liberal arts students. Preface. Recommended Reading Lists. Tables. Index. Numerous black-and-white figures. xvi + 641pp. 5⅜ × 8½.
24823-2 Pa. $11.95

28 SCIENCE FICTION STORIES, H. G. Wells. Novels, *Star Begotten* and *Men Like Gods,* plus 26 short stories: "Empire of the Ants," "A Story of the Stone Age," "The Stolen Bacillus," "In the Abyss," etc. 915pp. 5⅜ × 8½. (Available in U.S. only)
20265-8 Cloth. $10.95

HANDBOOK OF PICTORIAL SYMBOLS, Rudolph Modley. 3,250 signs and symbols, many systems in full; official or heavy commercial use. Arranged by subject. Most in Pictorial Archive series. 143pp. 8⅛ × 11. 23357-X Pa. $5.95

INCIDENTS OF TRAVEL IN YUCATAN, John L. Stephens. Classic (1843) exploration of jungles of Yucatan, looking for evidences of Maya civilization. Travel adventures, Mexican and Indian culture, etc. Total of 669pp. 5⅜ × 8½.
20926-1, 20927-X Pa., Two-vol. set $9.90

DEGAS: An Intimate Portrait, Ambroise Vollard. Charming, anecdotal memoir by famous art dealer of one of the greatest 19th-century French painters. 14 black-and-white illustrations. Introduction by Harold L. Van Doren. 96pp. 5⅜ × 8½.
25131-4 Pa. $3.95

PERSONAL NARRATIVE OF A PILGRIMAGE TO ALMANDINAH AND MECCAH, Richard Burton. Great travel classic by remarkably colorful personality. Burton, disguised as a Moroccan, visited sacred shrines of Islam, narrowly escaping death. 47 illustrations. 959pp. 5⅜ × 8½.   21217-3, 21218-1 Pa., Two-vol. set $17.90

PHRASE AND WORD ORIGINS, A. H. Holt. Entertaining, reliable, modern study of more than 1,200 colorful words, phrases, origins and histories. Much unexpected information. 254pp. 5⅜ × 8½.   20758-7 Pa. $5.95

THE RED THUMB MARK, R. Austin Freeman. In this first Dr. Thorndyke case, the great scientific detective draws fascinating conclusions from the nature of a single fingerprint. Exciting story, authentic science. 320pp. 5⅜ × 8½. (Available in U.S. only)   25210-8 Pa. $5.95

AN EGYPTIAN HIEROGLYPHIC DICTIONARY, E. A. Wallis Budge. Monumental work containing about 25,000 words or terms that occur in texts ranging from 3000 B.C. to 600 A.D. Each entry consists of a transliteration of the word, the word in hieroglyphs, and the meaning in English. 1,314pp. 6⅛ × 10.
23615-3, 23616-1 Pa., Two-vol. set $27.90

THE COMPLEAT STRATEGYST: Being a Primer on the Theory of Games of Strategy, J. D. Williams. Highly entertaining classic describes, with many illustrated examples, how to select best strategies in conflict situations. Prefaces. Appendices. xvi + 268pp. 5⅜ × 8½.   25101-2 Pa. $5.95

THE ROAD TO OZ, L. Frank Baum. Dorothy meets the Shaggy Man, little Button-Bright and the Rainbow's beautiful daughter in this delightful trip to the magical Land of Oz. 272pp. 5⅜ × 8.   25208-6 Pa. $4.95

POINT AND LINE TO PLANE, Wassily Kandinsky. Seminal exposition of role of point, line, other elements in non-objective painting. Essential to understanding 20th-century art. 127 illustrations. 192pp. 6½ × 9¼.   23808-3 Pa. $4.50

LADY ANNA, Anthony Trollope. Moving chronicle of Countess Lovel's bitter struggle to win for herself and daughter Anna their rightful rank and fortune— perhaps at cost of sanity itself. 384pp. 5⅜ × 8½.   24669-8 Pa. $6.95

EGYPTIAN MAGIC, E. A. Wallis Budge. Sums up all that is known about magic in Ancient Egypt: the role of magic in controlling the gods, powerful amulets that warded off evil spirits, scarabs of immortality, use of wax images, formulas and spells, the secret name, much more. 253pp. 5⅜ × 8½.   22681-6 Pa. $4.50

THE DANCE OF SIVA, Ananda Coomaraswamy. Preeminent authority unfolds the vast metaphysic of India: the revelation of her art, conception of the universe, social organization, etc. 27 reproductions of art masterpieces. 192pp. 5⅜ × 8½.
24817-8 Pa. $5.95

CHRISTMAS CUSTOMS AND TRADITIONS, Clement A. Miles. Origin, evolution, significance of religious, secular practices. Caroling, gifts, yule logs, much more. Full, scholarly yet fascinating; non-sectarian. 400pp. 5⅜ × 8½.
23354-5 Pa. $6.50

THE HUMAN FIGURE IN MOTION, Eadweard Muybridge. More than 4,500 stopped-action photos, in action series, showing undraped men, women, children jumping, lying down, throwing, sitting, wrestling, carrying, etc. 390pp. 7⅞ × 10⅝.
20204-6 Cloth. $19.95

THE MAN WHO WAS THURSDAY, Gilbert Keith Chesterton. Witty, fast-paced novel about a club of anarchists in turn-of-the-century London. Brilliant social, religious, philosophical speculations. 128pp. 5⅜ × 8½.
25121-7 Pa. $3.95

A CEZANNE SKETCHBOOK: Figures, Portraits, Landscapes and Still Lifes, Paul Cezanne. Great artist experiments with tonal effects, light, mass, other qualities in over 100 drawings. A revealing view of developing master painter, precursor of Cubism. 102 black-and-white illustrations. 144pp. 8¾ × 6⅝.
24790-2 Pa. $5.95

AN ENCYCLOPEDIA OF BATTLES: Accounts of Over 1,560 Battles from 1479 B.C. to the Present, David Eggenberger. Presents essential details of every major battle in recorded history, from the first battle of Megiddo in 1479 B.C. to Grenada in 1984. List of Battle Maps. New Appendix covering the years 1967–1984. Index. 99 illustrations. 544pp. 6½ × 9¼.
24913-1 Pa. $14.95

AN ETYMOLOGICAL DICTIONARY OF MODERN ENGLISH, Ernest Weekley. Richest, fullest work, by foremost British lexicographer. Detailed word histories. Inexhaustible. Total of 856pp. 6½ × 9¼.
21873-2, 21874-0 Pa., Two-vol. set $17.00

WEBSTER'S AMERICAN MILITARY BIOGRAPHIES, edited by Robert McHenry. Over 1,000 figures who shaped 3 centuries of American military history. Detailed biographies of Nathan Hale, Douglas MacArthur, Mary Hallaren, others. Chronologies of engagements, more. Introduction. Addenda. 1,033 entries in alphabetical order. xi + 548pp. 6½ × 9¼. (Available in U.S. only)
24758-9 Pa. $11.95

LIFE IN ANCIENT EGYPT, Adolf Erman. Detailed older account, with much not in more recent books: domestic life, religion, magic, medicine, commerce, and whatever else needed for complete picture. Many illustrations. 597pp. 5⅜ × 8½.
22632-8 Pa. $8.95

HISTORIC COSTUME IN PICTURES, Braun & Schneider. Over 1,450 costumed figures shown, covering a wide variety of peoples: kings, emperors, nobles, priests, servants, soldiers, scholars, townsfolk, peasants, merchants, courtiers, cavaliers, and more. 256pp. 8⅜ × 11¼.
23150-X Pa. $7.95

THE NOTEBOOKS OF LEONARDO DA VINCI, edited by J. P. Richter. Extracts from manuscripts reveal great genius; on painting, sculpture, anatomy, sciences, geography, etc. Both Italian and English. 186 ms. pages reproduced, plus 500 additional drawings, including studies for *Last Supper, Sforza* monument, etc. 860pp. 7⅞ × 10¾. (Available in U.S. only) 22572-0, 22573-9 Pa., Two-vol. set $25.90

THE ART NOUVEAU STYLE BOOK OF ALPHONSE MUCHA: All 72 Plates from "Documents Decoratifs" in Original Color, Alphonse Mucha. Rare copyright-free design portfolio by high priest of Art Nouveau. Jewelry, wallpaper, stained glass, furniture, figure studies, plant and animal motifs, etc. Only complete one-volume edition. 80pp. 9⅜ × 12¼. 24044-4 Pa. $8.95

ANIMALS: 1,419 COPYRIGHT-FREE ILLUSTRATIONS OF MAMMALS, BIRDS, FISH, INSECTS, ETC., edited by Jim Harter. Clear wood engravings present, in extremely lifelike poses, over 1,000 species of animals. One of the most extensive pictorial sourcebooks of its kind. Captions. Index. 284pp. 9 × 12.
23766-4 Pa. $9.95

OBELISTS FLY HIGH, C. Daly King. Masterpiece of American detective fiction, long out of print, involves murder on a 1935 transcontinental flight—"a very thrilling story"—NY Times. Unabridged and unaltered republication of the edition published by William Collins Sons & Co. Ltd., London, 1935. 288pp. 5⅜ × 8½. (Available in U.S. only) 25036-9 Pa. $4.95

VICTORIAN AND EDWARDIAN FASHION: A Photographic Survey, Alison Gernsheim. First fashion history completely illustrated by contemporary photographs. Full text plus 235 photos, 1840–1914, in which many celebrities appear. 240pp. 6½ × 9¼. 24205-6 Pa. $6.00

THE ART OF THE FRENCH ILLUSTRATED BOOK, 1700–1914, Gordon N. Ray. Over 630 superb book illustrations by Fragonard, Delacroix, Daumier, Doré, Grandville, Manet, Mucha, Steinlen, Toulouse-Lautrec and many others. Preface. Introduction. 633 halftones. Indices of artists, authors & titles, binders and provenances. Appendices. Bibliography. 608pp. 8⅜ × 11¼. 25086-5 Pa. $24.95

THE WONDERFUL WIZARD OF OZ, L. Frank Baum. Facsimile in full color of America's finest children's classic. 143 illustrations by W. W. Denslow. 267pp. 5⅜ × 8½. 20691-2 Pa. $5.95

FRONTIERS OF MODERN PHYSICS: New Perspectives on Cosmology, Relativity, Black Holes and Extraterrestrial Intelligence, Tony Rothman, et al. For the intelligent layman. Subjects include: cosmological models of the universe; black holes; the neutrino; the search for extraterrestrial intelligence. Introduction. 46 black-and-white illustrations. 192pp. 5⅜ × 8½. 24587-X Pa. $6.95

THE FRIENDLY STARS, Martha Evans Martin & Donald Howard Menzel. Classic text marshalls the stars together in an engaging, non-technical survey, presenting them as sources of beauty in night sky. 23 illustrations. Foreword. 2 star charts. Index. 147pp. 5⅜ × 8½. 21099-5 Pa. $3.50

FADS AND FALLACIES IN THE NAME OF SCIENCE, Martin Gardner. Fair, witty appraisal of cranks, quacks, and quackeries of science and pseudoscience: hollow earth, Velikovsky, orgone energy, Dianetics, flying saucers, Bridey Murphy, food and medical fads, etc. Revised, expanded In the Name of Science. "A very able and even-tempered presentation."—The New Yorker. 363pp. 5⅜ × 8.
20394-8 Pa. $6.50

ANCIENT EGYPT: ITS CULTURE AND HISTORY, J. E Manchip White. From pre-dynastics through Ptolemies: society, history, political structure, religion, daily life, literature, cultural heritage. 48 plates. 217pp. 5⅜ × 8½. 22548-8 Pa. $4.95

SIR HARRY HOTSPUR OF HUMBLETHWAITE, Anthony Trollope. Incisive, unconventional psychological study of a conflict between a wealthy baronet, his idealistic daughter, and their scapegrace cousin. The 1870 novel in its first inexpensive edition in years. 250pp. 5⅜ × 8½. 24953-0 Pa. $5.95

LASERS AND HOLOGRAPHY, Winston E. Kock. Sound introduction to burgeoning field, expanded (1981) for second edition. Wave patterns, coherence, lasers, diffraction, zone plates, properties of holograms, recent advances. 84 illustrations. 160pp. 5⅜ × 8¼. (Except in United Kingdom) 24041-X Pa. $3.50

INTRODUCTION TO ARTIFICIAL INTELLIGENCE: SECOND, EN-LARGED EDITION, Philip C. Jackson, Jr. Comprehensive survey of artificial intelligence—the study of how machines (computers) can be made to act intelli-gently. Includes introductory and advanced material. Extensive notes updating the main text. 132 black-and-white illustrations. 512pp. 5⅜ × 8½. 24864-X Pa. $8.95

HISTORY OF INDIAN AND INDONESIAN ART, Ananda K. Coomaraswamy. Over 400 illustrations illuminate classic study of Indian art from earliest Harappa finds to early 20th century. Provides philosophical, religious and social insights. 304pp. 6⅜ × 9⅜. 25005-9 Pa. $8.95

THE GOLEM, Gustav Meyrink. Most famous supernatural novel in modern European literature, set in Ghetto of Old Prague around 1890. Compelling story of mystical experiences, strange transformations, profound terror. 13 black-and-white illustrations. 224pp. 5⅜ × 8½. (Available in U.S. only) 25025-3 Pa. $5.95

ARMADALE, Wilkie Collins. Third great mystery novel by the author of *The Woman in White* and *The Moonstone*. Original magazine version with 40 illustrations. 597pp. 5⅜ × 8½. 23429-0 Pa. $9.95

PICTORIAL ENCYCLOPEDIA OF HISTORIC ARCHITECTURAL PLANS, DETAILS AND ELEMENTS: With 1,880 Line Drawings of Arches, Domes, Doorways, Facades, Gables, Windows, etc., John Theodore Haneman. Sourcebook of inspiration for architects, designers, others. Bibliography. Captions. 141pp. 9 × 12. 24605-1 Pa. $6.95

BENCHLEY LOST AND FOUND, Robert Benchley. Finest humor from early 30's, about pet peeves, child psychologists, post office and others. Mostly unavailable elsewhere. 73 illustrations by Peter Arno and others. 183pp. 5⅜ × 8½. 22410-4 Pa. $3.95

ERTÉ GRAPHICS, Erté. Collection of striking color graphics: *Seasons, Alphabet, Numerals, Aces* and *Precious Stones*. 50 plates, including 4 on covers. 48pp. 9⅜ × 12¼. 23580-7 Pa. $6.95

THE JOURNAL OF HENRY D. THOREAU, edited by Bradford Torrey, F. H. Allen. Complete reprinting of 14 volumes, 1837–61, over two million words; the sourcebooks for *Walden*, etc. Definitive. All original sketches, plus 75 photographs. 1,804pp. 8½ × 12¼. 20312-3, 20313-1 Cloth., Two-vol. set $80.00

CASTLES: THEIR CONSTRUCTION AND HISTORY, Sidney Toy. Traces castle development from ancient roots. Nearly 200 photographs and drawings illustrate moats, keeps, baileys, many other features. Caernarvon, Dover Castles, Hadrian's Wall, Tower of London, dozens more. 256pp. 5⅜ × 8¼. 24898-4 Pa. $5.95

AMERICAN CLIPPER SHIPS: 1833–1858, Octavius T. Howe & Frederick C. Matthews. Fully-illustrated, encyclopedic review of 352 clipper ships from the period of America's greatest maritime supremacy. Introduction. 109 halftones. 5 black-and-white line illustrations. Index. Total of 928pp. 5⅜ × 8½.
25115-2, 25116-0 Pa., Two-vol. set $17.90

TOWARDS A NEW ARCHITECTURE, Le Corbusier. Pioneering manifesto by great architect, near legendary founder of "International School." Technical and aesthetic theories, views on industry, economics, relation of form to function, "mass-production spirit," much more. Profusely illustrated. Unabridged translation of 13th French edition. Introduction by Frederick Etchells. 320pp. 6⅛ × 9¼. (Available in U.S. only)
25023-7 Pa. $8.95

THE BOOK OF KELLS, edited by Blanche Cirker. Inexpensive collection of 32 full-color, full-page plates from the greatest illuminated manuscript of the Middle Ages, painstakingly reproduced from rare facsimile edition. Publisher's Note. Captions. 32pp. 9⅜ × 12¼.
24345-1 Pa. $4.95

BEST SCIENCE FICTION STORIES OF H. G. WELLS, H. G. Wells. Full novel *The Invisible Man*, plus 17 short stories: "The Crystal Egg," "Aepyornis Island," "The Strange Orchid," etc. 303pp. 5⅜ × 8½. (Available in U.S. only)
21531-8 Pa. $4.95

AMERICAN SAILING SHIPS: Their Plans and History, Charles G. Davis. Photos, construction details of schooners, frigates, clippers, other sailcraft of 18th to early 20th centuries—plus entertaining discourse on design, rigging, nautical lore, much more. 137 black-and-white illustrations. 240pp. 6⅛ × 9¼.
24658-2 Pa. $5.95

ENTERTAINING MATHEMATICAL PUZZLES, Martin Gardner. Selection of author's favorite conundrums involving arithmetic, money, speed, etc., with lively commentary. Complete solutions. 112pp. 5⅜ × 8½.
25211-6 Pa. $2.95

THE WILL TO BELIEVE, HUMAN IMMORTALITY, William James. Two books bound together. Effect of irrational on logical, and arguments for human immortality. 402pp. 5⅜ × 8½.
20291-7 Pa. $7.50

THE HAUNTED MONASTERY and THE CHINESE MAZE MURDERS, Robert Van Gulik. 2 full novels by Van Gulik continue adventures of Judge Dee and his companions. An evil Taoist monastery, seemingly supernatural events; overgrown topiary maze that hides strange crimes. Set in 7th-century China. 27 illustrations. 328pp. 5⅜ × 8½.
23502-5 Pa. $5.95

CELEBRATED CASES OF JUDGE DEE (DEE GOONG AN), translated by Robert Van Gulik. Authentic 18th-century Chinese detective novel; Dee and associates solve three interlocked cases. Led to Van Gulik's own stories with same characters. Extensive introduction. 9 illustrations. 237pp. 5⅜ × 8½.
23337-5 Pa. $4.95